Fermentation Factor

How to Reap the Amazing Health Benefits of Traditionally Fermented Foods

by
Abigail Adams

Fermentation Factor
© 2012 Abigail Adams

A product of Solutions From Science

Notice of Rights
Manufactured in the United States of America. All rights reserved. No part of this book may be reproduced in any form or by any electronic or mechanical means, including information storage or retrieval systems, without permission in writing by the copyright owner. For more products by Solutions From Science, please visit us on the web at www. solutionsfromscience.com

Photographs used from Wikimedia Commons are used under the Creative Commons Attribution-ShareAlike 3.0 Unported License. Author attribution is given with each photograph.

Notice of Liability
The information in this book is distributed on an "as is" basis, for informational purposes only, without warranty. While every precaution has been taken in the production of this book, neither the copyright owner nor Heritage Press Publications, LLC shall have any liability to any person or entity with respect to any liability, loss, or damage caused or alleged to be caused directly or indirectly by the instructions contained in this book.

Published by:
Heritage Press Publications, LLC
PO Box 561
Collinsville, MS 39325

ISBN-10: 1937660125
ISBN-13: 978-1-937660-12-3

Contents

Fermentation Factor
An Introduction to Fermenting   *5*

Tried and True
Tools, Tips, and Techniques for Fermenting   *9*

Back to Basics
Fermented Fruits and Veggies   *16*

Sweet and Sour
Fermented Breads and Grains   *36*

Delicious Drinks
Fermented Beverages   *55*

Curds and Whey
Fermenting Dairy   *64*

This and That
Fermented or Not?   *71*

How to Grow Your Own Food to Ferment   *80*

Appendix
Additional Resources   *121*

CHAPTER I

Fermentation Factor
An Introduction to Fermenting

"Cabbage could be in medical science what bread is in nutrition; cabbage is the physician of the poor."
-Dr. Blanc, Parisian physician, 1881

What if we told you that there was a way to preserve your vegetables for months without using heat or cold? What if we told you that you that this method preserves the original freshness and nutritional value of your produce without using potentially harmful chemicals or any other substances that most of us have difficulty pronouncing or spelling?

This is not a new scientific breakthrough discovered by a dedicated team of lab wizards working in sterile white coats. This knowledge is older than the pyramids, yet modern science has shown that it helps reverse some of the ill effects of our pre-packaged, over-processed lifestyle.

If you've read the cover of this book, you know we're talking about fermentation.

Fermentation Techniques are Older Than Dirt

Historians and archeologists tell us that humans discovered the benefits of fermentation right around the time they moved away from hunter-gatherer societies. The development of agriculture caused tribes and primitive villages to harvest more food than they could consume. Since crops began to spoil soon after they were harvested and the distance between settlements was too large for trading fresh food items, it was essential to discover ways to store food for later use. As people began to harness the natural ability of fermentation to keep their food from spoiling, a new category of food was born.

Fermentation Factor

Chinese records show that humans in that area of the world discovered how to ferment cabbage over 6,000 years ago. A dish of fermented vegetables similar to kimchi is said to be an important part of the diet of the workers who built the Great Wall of China. When Mongol armies finally got around that wall and conquered China, they adopted a variation of this durable fermented vegetable creation as one of their main travel foods. Genghis Khan's troops conquered a large portion of the known world with fermented vegetables as one of their staple foods.

Centuries later, Captain James Cook of the British Navy developed a reputation for maintaining a healthy crew on long sea voyages. His secret was sauerkraut. High in vitamin C, fermented cabbage protected his sailors from scurvy. Cook's confidence in sauerkraut was so strong that he ordered 25,000 pounds of it to outfit two ships before a long journey.

Throughout history, virtually every civilization has experienced the numerous benefits of preserving their food by fermentation. And again, it was discovered long before our society started using chemicals and lab engineering to process food.

Fermentation is Natural

Fermentation is the ever-present interaction between bacteria and fungi and the sugars, starches, and proteins in food. During the fermentation process, carbohydrates turn to alcohols or acids as yeasts or bacteria digest them.

The type of fermentation that produces most of the food we eat is lactic acid fermentation, produced by *Lactobacilli*. These beneficial bacteria are naturally present on almost all foods— and in the human body. That's right—you probably have millions of *Lactobacilli* inside you right now, helping to maintain ideal conditions inside your intestines. These good bacteria actually help fight the bacteria that make you sick. In food, the acidic conditions created by *Lactobacilli* activity are what keep food from spoiling. Yogurt, sauerkraut, and sour pickles are all the products of lactic acid fermentation.

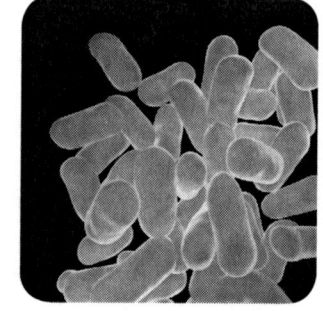

Health Benefits of Fermented Foods

Live fermented dishes introduce good bacteria into your digestive system. These probiotic bacteria strengthen your immune system, helping your body fight off diseases. As an added bonus, this substantial quantity of good bacteria also inhibits the growth of harmful microorganisms that can cause diarrhea and other digestive complaints.

Because fermenting foods at home incorporates the flora, fungi, and bacteria in the world around you, eating fermented foods helps your body to become acclimated to the area in which you live. The naturally occurring bacteria in your area build your immunity even more than eating purchased fermented foods—even foods that are fermented using live bacteria. By eating home-fermented foods, you are essentially becoming one with your environment. This leaves you less susceptible to harmful outside inputs.

Higher Nutrient Density

Because fermentation doesn't involve using extreme heat or extreme cold, the nutrients in your vegetables are preserved. Not only are they preserved, but they also become more readily available to your body.

Comparing unfermented and fermented mixtures of barley, lentils, milk powder, and tomato pulp, a study published in 1997 in the journal Nutritional Health (11(3); pp. 139-147) discovered that "starch digestibility almost doubled in the fermented mixture." In 1999, the United Nations Food and Agriculture Organization reported in their agriculture service bulletin *Fermented Foods: A Global Perspective,* that "fermentation improves mineral bioavailability in foods." What that means in common terms is that your body gets more of the useful minerals in your veggies when you ferment them.

Fermentation not only makes more of the nutrients in your food available to you, it also creates new nutrients. As the microbial cultures that make fermentation possible go through their life cycle, they create essential B vitamins, including folic acid, riboflavin, niacin, thiamine, and biotin. *Lactobacilli* also create omega-3 fatty acids, which are important in cell membrane and immune system functions.

In short, eating live, fermented foods is a good idea if you want a healthy body.

But Why Ferment Your Own?

At this point, some of you might be wondering why you need to learn how to ferment your own food. After all, your local supermarket carries sauerkraut, pickled cucumbers and beets, black olives, and might even have a form of kimchi. Wouldn't it be easier just to purchase what you need to get many of the same health benefits until you learn more about becoming self-sufficient?

The short answer to that question is no.

Modern food-processing methods have rendered most of the fermented foods found in supermarkets virtually sterile; in fact, many traditionally fermented foods you'll find on your grocer's shelves aren't even fermented these days.

For instance, black olives that were traditionally created using fermentation are now simply treated with lye to remove the bitterness and then packed in a solution of lactic acid, acetic acid, sodium benzoate, and potassium sorbate.

Photo courtesy of Wikimedia Commons by Albert Cahalan, released to the public domain.

Many commercial sauerkrauts and pickles are now created using vinegar. While the vinegar may (or may not) be created using natural fermentation, the actual vegetable will have barely a hint of the nutritional value of one that you ferment at home.

The few foods you find at your supermarket that are actually made using a natural fermentation method are usually stripped of their healthy properties by the use of high-heat pasteurization, which kills off all the lactic acid-producing bacteria and ruins the chance that this food will confer any of the intestinal and overall health benefits mentioned above.

You can fight back, and you can protect the health of your family by learning how to ferment your own organically grown vegetables. It is simple and often fun. The time has come to abandon the consumption of artificial preservatives with names you can't pronounce and return to the natural and healthy ways—not only to preserve your food but to increase its nutritional value. We congratulate you for taking the first step, and we hope you enjoy the journey.

CHAPTER 2

Tried and True
Tools, Tips, and Techniques for Fermenting

Fermentation is easy; fermentation is also hard. When you come right down to it, you're trying to create the proper conditions for organisms that you can't even see to create a desired result. When things go wrong, a fermentation project can seem an awful lot like trying to herd invisible kittens.

Despite your best efforts, you are never completely in control of the forces that make fermentation possible. There will be times when your room is too hot, too cold, or too humid for the microorganisms to do their work effectively.

While we have taken every effort to anticipate problems and to show you how to avoid them, there will come a day when you lift the cover of your container and catch a whiff of something nasty. Don't get discouraged; add this material to your compost heap or throw it out. Odds are very good that your next batch will turn out fine.

One of the great pleasures of preserving your food by fermentation is the unique flavors you will encounter every time. Most of these experiences will be surprisingly pleasant. Since the results vary according to the quality of your ingredients, the season, overall environment, heat, humidity, and many other factors, each time you make the same recipe, it will probably taste a little bit different.

Don't be afraid of the uncertainty factor. You are working with methods that were discovered long before bland, standardized food took over the highways and byways of the world. You are creating living food that will nourish your body in ways that sterile, pre-packaged, always-the-same products cannot. By fermenting your own food, you will now have naturally preserved food that will boost your immune system, aid your digestion, and provide you with beneficial vitamins and minerals.

Preservation Without Power

Although natural fermentation is a way to preserve food and retain (or even enhance) the nutrients within that food, you will still encounter issues when it comes to storing fermented dishes. Let's face it: nothing lasts forever. Fermentation can preserve the viability of your food, but it does have limitations. Up to a certain point, time will enhance the flavors and textures of your ferment. After that point has passed, the flavors, textures, and nutritional value will diminish.

As a general rule, fermentation happens faster in warmer temperatures. After your fermented food has come close to the flavor and texture you desire, storing it in the fridge, a cool root cellar, or a basement will slow down the fermentation process and extend the life of the food. Foods fermented in brine will generally survive longer in a saltier brine, though you may have to rinse them several times to cut back on the salty taste. Many experienced fermenters prefer to ferment their food in cool cellars or basements. Foods fermented in a cool environment take more time to achieve the ideal taste and texture, but they tend to be crisper and survive longer.

It is a good idea to remember that while the cooler temperature of your fridge will slow down the fermentation process, it does not stop it. When storing fermented foods in your refrigerator, you'll still want to avoid tightly closed caps. The gasses produced by fermentation can explode any tightly contained containers that you might neglect. Check your stored food regularly to avoid unpleasant surprises.

Following proper fermentation techniques should help you and your family to extend your harvest and gain the ability to enjoy nutritional food year-round. Your biggest investments will be time, patience, and the willingness to learn from your mistakes. As you experience the marvels of fermentation, you will learn to appreciate the complex wonders of God's creation and to understand how some of the smallest pieces of that creation can help sustain you and your family through times of plenty and times of famine.

Fermentation Equipment

If you are on a tight budget, one of the best things about preserving food through fermentation is that you don't need to invest a fortune in special

Fermentation Factor

equipment or pay out a whole lot of extra money to your utility company for energy-intense cooking or freezing processes.

As an added benefit, fermentation does not require a major investment in fancy equipment; in fact, you probably have most of the items readily available in your kitchen.

Basic supplies:
- good ceramic crock or food-grade plastic bucket
- wide-mouth glass jars
- small glass jars
- glass gallon jug
- sturdy plates of various sizes
- ceramic bowls
- spoons
- cheesecloth or clean towels
- rubber bands

If you have these items on hand, you can take on virtually any simple fermentation project. As you move through more complex fermentation techniques, you might want to invest in a few special ceramic containers, a grain grinder, and a few other items, but when you are just getting started, we suggest keeping it simple. It is always a good idea to remember that most fermentation processes were invented hundreds, if not thousands, of years before humans had the technology to make them more complicated. When in doubt, keep it simple and remember that your ancestors had no choice. They had no complicated options, yet they managed to create many of the fermented delights that we still enjoy today.

Picture of a Harsch crock with ceramic weights for fermenting foods. Photo courtesy of Wikimedia Commons by Bdubay

Fermentation Factor

If, however, you cannot resist the urge to invest in a little high-tech equipment, items like the Pick-It can be very useful for combining several of these tools into one piece of equipment. More information can be found in chapter 8 under Additional Resources.

Fermentation Basics and Troubleshooting

You can preserve virtually any vegetable you can grow in your garden by fermentation. Some people prefer simple, single-vegetable preservation, while others enjoy mixing up a batch of vegetables and spices with a wide variety of complex flavors and textures.

Either way, you'll enjoy the power-packed benefits of natural fermentation without having to memorize a whole lot of complex formulas. For the most part, vegetable fermentation is simple and fun. Of course, nothing is foolproof in this world. An ounce or two of knowledge can go a long way in avoiding pounds of frustration.

The most common mistakes people bump into when tackling their first batch of fermented vegetables include using the wrong container, failing to keep their mixture pressed under the brine, or using the wrong ingredients.

The biggest mistake you can make is attempting to ferment in a metal pot. Metal will react with the salt you'll be using. Once your ferment gets going, the metal will also be leached by the acids produced by the fermentation process. These dual dynamics will not only ruin the flavor of your preserved vegetables, but will potentially create dangerous compounds. To make a long lecture a bit shorter, do NOT use metal: no copper, no iron, no aluminum, no stainless steel, no pewter, no titanium, no super-duper whiz-bang alloy that the smiling faces on your favorite shopping network have claimed are safe for all purposes.

Do not use any metal tools in the fermentation process.
Picture courtesy of Wikimedia Commons by FASTILY

Metal and fermentation do not mix—period.

Another mistake is trying to get away with using a metal pot by lining it with a trash bag or other liner. These materials are not food grade and are at risk for tearing, which brings you back to the original problem of the metal containers.

Traditionally, fermented foods have been made in gourds, animal membranes, and ceramic containers. A large ceramic crock is an ideal choice, but they are often expensive and increasingly hard to find.

Wide-mouth glass jars are favored by some long-time fermenters. You'll need a glass jar big enough to insert a plate and a weight, so we're talking a jar of some size here. For smaller batches, you can use a wide-mouth quart jar and a smaller jelly jar that will fit inside the opening. Another option that is becoming increasingly popular is the five-gallon plastic bucket that you can often find at a delicatessen. When using plastic, always make sure to use food-grade plastic.

Whichever appropriate container you choose, you'll need a way to press down your mixture. Many fermenters have had good success using a plate small enough to fit into the container but wide enough to cover most of the veggie mix. A gallon water jug makes a great weight to press everything down, though there are people who use cleaned and sterilized rocks or sealed plastic bags filled with brine.

The purpose of these items is to keep your fermenting vegetables under the brine. Your brine will protect your ferment from picking up mold or other substances that can ruin your preservation attempts. As long as you keep everything under the brine, any bloom (the white layer) you may find floating on the top is harmless. It is a natural surface phenomenon and nothing to worry about; your developing ferment is safe under the anaerobic protection of the brine. Just scoop it off and don't worry. Your delicious flavors are safe in the salt and developing acidity of the briny liquid. However, if you fail to keep everything below the brine line, you will soon have a problem on your hands, and you'll probably just create a slimy mess best suited for the compost heap.

There is some debate about the use of chlorinated water in fermentation. Some claim that fermentation with chlorinated water will never succeed, while others say there is no problem using the water from your tap. Here's the catch: chlorine is added to water systems to kill microorganisms—some of the very same critters that are going to make the whole

Fermentation Factor

fermentation process happen. Fermented food is living food. If you want that life to prosper, you have to make sure that chlorine is not going to interfere. However, different water systems have different amounts of chlorine; well water generally has none. It's possible that you won't have any trouble with your water—if your water smells and tastes fine, don't be afraid to give it a try. If you have difficulties with your fermentation, try boiling your tap water for a few minutes and let it cool before starting your fermentation project. This should remove enough chlorine to give those necessary microorganisms a chance to do their job.

The next ingredient you'll need to consider is the salt. Standard supermarket table salt contains iodine and anti-caking agents. Like chlorine, iodine can destroy the very microorganisms you need for the miracle of fermentation to take place, and anti-caking agents can make your brine become cloudy. To avoid disappointment, you should use sea salt or pickling salt. Coarse kosher salt also works, but due to the larger grains your measurements will vary. As a rule of thumb, figure a tablespoon and a half of coarse kosher salt for every tablespoon of finer-grain salt when doing your conversions.

Salt is an important part of the fermenting process, so avoid the temptation to skimp or leave out salt altogether. If you find that your pickles are coming out a bit saltier than you prefer, give them a good rinse or soak before eating them. If salt content is a health concern, see chapter 7 for lower-sodium recipes.

Finally, you need to consider the produce you are fermenting. While it is possible to ferment older produce, you will experience some problems. Time strips the natural juices and nutrients from most vegetables. The flavor, texture, and health value of your finished product will all be impacted. Frankly, starting with fresh, organic fruits and vegetables is your best bet, particularly when you are just starting out. You also want to avoid produce with any blemishes.

It is important to use organic produce for fermenting. Most of the produce you see at your local supermarket is coated with a thin layer of wax. This helps seal in the moisture to help them survive their long voyage from the factory farms to your produce section. Pesticides and other unhealthy substances often lurk under this wax, which you just can't wash off.

Fermentation Factor

Trying to ferment a waxed veggie is a lot like trying to ferment a candle. Don't bother.

If you're not growing your own fruits and veggies, make sure you choose unwaxed, organic produce. Additionally, when you use any additional ingredients (like spices), you will want to make sure that they do not contain any added preservatives or artificial ingredients. Fermentation is a complex process; you don't want to make it any more complicated by adding in unnecessary chemicals.

Storage

After the initial fermentation period is complete, you will need to store your fermented goodies somewhere. Cool storage (like a cellar or refrigerator) will slow down fermentation, but it won't completely stop it. Keep this in mind and keep lids loose, otherwise you might end up with a mess on your hands. Also, because the fermentation process continues, flavors will continue to develop while in store. Don't be surprised if your pickled treats taste a bit strong after a few weeks in the fridge; however, if you smell or taste anything nasty, you should discard the rest immediately.

Photo courtesy of Wikimedia Commons by telethon

You can also can your pickles if you need to store them for a longer period of time or if you don't have enough room in your cool storage, but remember that the heat from canning will destroy some of the nutrients and living organisms that make fermentation so beneficial. However, if this is a step you want or need to take, it is easily achieved through a boiling-water bath. Check your Ball Blue Book or other canning guide for the steps you need to take to process your pickles safely.

Now that you've got your tools and ingredients, you're ready to get fermenting!

CHAPTER 3

BACK TO BASICS
Fermented Fruits and Veggies

Fermenting Vegetables

The easiest and most common fermented foods to make are fermented vegetables. From pickles to sauerkraut and other vegetable combinations, fermenting vegetables from the garden is one way to preserve food without power. This chapter provides recipes for common vegetable ferments.

Simple Homemade Pickling Spice

If you're determined to use a commercial pickling spice combo when you pickle by fermentation, remember to read the ingredients carefully to make sure there are no preservatives or artificial ingredients. You never want to add substances that will inhibit your ferment.

Your best choice is always to make your own pickling spice. You can customize the flavors to your liking, and you can ensure that no unwanted chemicals enter your ferment. We've included a recipe here, but there are unlimited varieties out there. Feel free to update this recipe to your taste preferences, or search cookbooks and online resources for more options.

This is a fairly spicy mix filled with delightful flavors Unlike most projects in this book, creating this pickling spice blend does involve a small amount of cooking over low heat. The purpose of this step is to release additional flavor from otherwise reluctant seeds. You can skip this step if you wish, but you'll be missing out on a little extra zest if you do.

Equipment:
- Small pan
- Small mixing bowl
- Airtight storage container

Fermentation Factor

- Ingredients:
- 2 tablespoons black peppercorns
- 2 tablespoons coriander
- 2 tablespoons mustard seeds
- 1 medium cinnamon stick
- 18 bay leaves
- 1 tablespoon ground mace
- 2 tablespoons allspice berries
- 2 tablespoons whole cloves
- 1 tablespoon ground ginger
- 1 tablespoon hot red pepper flakes

1. Put the peppercorns, coriander, and mustard seeds in a dry pan and heat them over low heat, stirring constantly. Ideally, the heat should be so low that the seeds don't pop, but keep a lid handy just in case. All you're doing in this step is heating the seeds until they are fragrant. When your mixture starts to smell good, take it off the heat. Don't overcook or burn the seeds.

2. Use the back of a spoon, the flat side of a butter knife, or any other kitchen tool you have handy to crack your slightly heated seeds. When this task is done, put the cracked seeds in a bowl.

3. Either break the cinnamon stick into small pieces or crush it and add it to the bowl.

4. Crumple or tear up the bay leaves and add them to the bowl. Stir this mixture up a bit.

5. Add the rest of the ingredients and stir it all up again.

And that's all there is to it!

This recipe should yield about a cup of spicy goodness. Store your homemade pickling spice in a tightly

Fermentation Factor

sealed plastic bag or glass container. Be sure to label it so other folks in house know what it is. The longer you store the spice, the more it will lose its flavor, so try to use it as quickly as possible.

Pickles

The quest for crunchy, mouth-pleasing, old-fashioned sour pickles has driven many first-time fermenters to the brink of despair. Their dreams are dashed when their work results in a batch of soggy pickles or, worse, rotten partially pickled cucumbers. There are several ways in which pickling using basic fermentation techniques can go awry. This bad news, however, should not scare you away from fermenting pickles. All of the potential problems of pickling are avoidable if you take a little bit of extra care.

First of all, make sure you have read the tips in chapter 2 thoroughly. Remember that you need to carefully choose your tools and ingredients if you want successful pickles.

The next important factor in successful cucumber pickles is the brine. The strength of the brine is a rather controversial subject for some fermenters. Some old recipes call for enough salt to float an egg. That's about a 10% salt solution. While this density does lead to long-lasting sour pickles, it also makes them very salty, requiring a good, long cold-water soak before they're fit for eating.

Fermentation expert Sandor Ellix Katz suggests that a 5.4% salt brine is about average (Wild Fermentation, Chelsea Green Publishing Company, pp. 50-52). This is a good number for people who like really sour pickles, but it can be a bit intense for people who prefer pickles with less pucker.

As a rule of thumb, we usually recommend about 2 to 2½ tablespoons of salt per quart of water. This brings your brine into the range of 3.6% to 4.5%, which is enough to do the job without stressing your taste buds. If you find you prefer a little more pucker, try adding more salt in small amounts.

Room temperature is another factor to consider when making pickles. Fermentation takes place faster in warmer weather. If your room is warm, you need to keep a closer watch on your pickles, and you might have to pop them into the fridge sooner to slow the pickling process.

Fermentation Factor

Some individuals who have great success fermenting chopped vegetables have trouble creating pickles because they fail to keep the cucumbers below the brine. Unlike chopped veggies, which tend to lie flat, cucumbers can be a bit unruly. Any cucumbers that pop above your brine level will rot. Keep a plate and a weight on your developing pickles when you're not checking or tasting them. If you see any bloom developing, make sure you wash it off the plate and weight after skimming it off the brine surface.

After carefully following all of these steps, it is still possible to create a soft, mushy batch of pickles if you fail to bring some tannin into the mix.

Believe it or not, the best way to do this is with leaves. Adding a fresh grape leaf, horseradish leaf, or oak leaf to your container of fermenting cucumbers can go a long way toward giving them the crunchy texture you crave.

Now that we've made the whole concept of pickling sound like rocket science, you'll be amazed how simple the following cucumber pickles recipe is to follow.

Equipment:
- Large ceramic crock, wide-mouth glass container, or food-grade plastic bucket
- Good solid plate that fits inside your container
- Gallon jug filled with water or another weight
- Clean cloth napkin or small towel

Ingredients:
- 3 to 4 pounds small- to medium-sized cucumbers
- 3 quarts water (1 of them cold)
- 4 to 5 tablespoons salt (use 6 to 7 tablespoons if you chose coarse kosher salt)
- 5 or 6 cloves fresh garlic
- 3 tablespoons pickling spice
- 2 big bunches of organic dill with flowering heads (if fresh dill is not available, you can substitute 2 tablespoons of dried dill)
- 2 or 3 fresh grape, horseradish, or oak leaves

Fermentation Factor

Our experience has shown that it is best to soak your cucumbers in about a quart of cold water for about two hours before getting started. This soak will freshen up the cukes and give them a bit more snap.

1. Once your cucumbers have finished soaking, it is time to start converting them into pickles.

2. Pour the remaining 2 quarts of water into the clean container and add the salt. Stir it vigorously to dissolve the salt.

3. Chop up the cloves of garlic and add them to the container.

4. Stir in the pickling spices and dill.

5. Add the fresh leaves of your choice (grape, horseradish or oak) and stir again.

6. Add the cucumbers to the container. Push them under the brine with the plate and weigh the plate down. Double-check to be sure none of those unruly cukes have slipped out from under the weighted plate.

7. Cover the container with a cloth napkin or small towel and place it in a convenient corner of your kitchen. A cool spot is nice if you can find one, but don't worry too much if you can't. Although your fridge is probably quite cool, we're not ready for that step yet.

8. Check your container daily. If some bloom forms on the top of the brine, just skim off as much as you can and wash your plate and your weight.

9. After 3 or 4 days of fermenting, you'll want to start tasting your pickles. As they approach your desired degree of sour, it is time to start thinking about moving them to the fridge. Before moving them to the fridge, you may want to store them in smaller glass containers. This is fine, as long as you make sure that the brine continues to cover the pickles and you don't screw the lids down tightly. The fermentation process will slow in the cooler temperature of your fridge, but it will not stop completely. If you have tight lids, gasses may build up to the point where your jars burst. Loose lids can save you the messy process of cleaning up pickle juice.

Now that you know how to pickle cucumbers, you're ready to pickle other fresh produce from your garden. The process remains the same for all veggies, though you might want to cut down the amount of dill or even leave it out completely for other vegetables. Don't be afraid to try other options like beets, carrots, turnips, peppers, and squash. Your imagination really is the limit when it comes to your options here.

Variation: Sweet Pickles

In addition to sour pickles, you can also make delicious sweet pickles using fermentation. This recipe gives you a flavor similar to bread and butter pickles, but with a lot more nutrients than the ones you find on the shelves of your local supermarket.

Equipment:
- Small pan for heating spice mixture
- 1 quart wide-mouth glass jar (or other fermenting container)
- Smaller jar filled with water (should fit inside the quart jar)
- Clean towel or cloth

Ingredients:
- 1 pound cucumbers
- ½ cup mild onion, thinly sliced
- 1 small bay leaf
- 2 tablespoons fresh lemon juice
- 4 tablespoons whey (see chapter 6 for how to obtain your own whey)
- 2 quarts water

Spice mix:
- 2/3 cup sugar or maple syrup (if you choose sugar, you will need to add a bit of water to the mixture in order to make it a liquid)
- 2 teaspoons ground cinnamon, or use a small piece of cinnamon stick crumbled
- ½ teaspoon whole all-spice berries
- 5 whole cloves
- pinch of ground ginger
- pinch of dill seed

- 1 teaspoon mustard seed
- ¼ teaspoon celery seed
- ⅛ teaspoon red pepper flakes
- 2 teaspoons salt

As you did for sour pickles, you will want to soak your cucumbers in cold water for several hours before pickling. While they are soaking, you can prepare the rest of your ingredients.

1. Bring all the ingredients for the spice mix to a boil over medium heat. Immediately remove from heat and let spices steep until liquid has cooled to room temperature.

2. Cut cucumbers into ¼-inch slices. Alternate pickle slices and onion inside the 1 quart wide-mouth jar, laying them as flat as possible. Halfway through, place the bay leaf in the middle of the pickles. Continue filling the jar with cucumbers and onions.

3. When the spice mixture is cooled, stir in the lemon juice and whey. Pour this mixture over the cucumber slices in the jar, adding additional water to cover pickles as needed. Be sure to leave some space at the top of the jar.

4. Gently stir to mix any water you added with the spice mixture.

5. Place the smaller jar inside the mouth of the quart jar. This should push the pickle slices safely below the brine. Place in a cool place and allow to ferment at room temperature for 2 days.

6. After 2 days, your sweet pickles can be transferred to cold storage. Cover them loosely, as fermentation will slow but not completely stop in the cooler temperatures.

Simple Sauerkraut

Cabbage has long been a staple of many western diets. To preserve it between growing seasons, families converted a portion of their harvest into savory sauerkraut. Americans found many uses for this fermented delight. They put it on hot dogs; they combined it with corned beef, rye bread, Swiss cheese, and a dash of Thousand Island dressing to develop the classic Reuben sandwich; and they devoured it as a side dish with roast pork and other meals.

Since this tasty treat has so many fans and is so simple to make, it comes as no surprise that it is often one of the first fermented vegetable creations many people attempt.

Equipment:

- Large bowl
- Ceramic crock or other fermenting container with a capacity of a gallon or more
- Good solid plate that fits inside your container
- Your choice of weight
- Pillowcase or towel to use as a cloth cover

Ingredients:

- 5 pounds cabbage
- 3 tablespoons salt
- Spices or other ingredients as desired

Before you start, you will need to prepare your cabbage. You can choose to chop it into fine ribbons, a coarse cut, or large chunks—whichever works for you. You can use just the leaves or include the heart of the cabbage. Some people like to mix green cabbage with red cabbage to create eye-catching bright pink sauerkraut. Choose whichever you like best.

1. Layer the cabbage and salt in a large, clean bowl. This salt will pull water out of the cabbage and create your brine.

2. Mix in any spices you would like. Classic sauerkraut spices include celery seeds, dill seeds, juniper berries, and caraway seeds. If this is your first batch of kraut, try adding spices with a light hand so that you have baseline when your cabbage is fermented. On your next batch, you can adjust quantities and ingredients to taste. Many people also enjoy adding various vegetables and fruits to their sauerkraut mix. If this is your first batch, you may want to avoid the temptation of these additions until you've had a chance to taste the finished product of your first efforts. You'll have plenty of time to experiment later.

Fermentation Factor

3. Using a small amount at a time, place the salted cabbage in the fermenting container and press it down hard. Many fermenters use a fist for this task, while others reach for the tamping tool they prefer. Either way, what you're trying to do at this stage is force some water out of the cabbage. Continue this packing-and-pressing process until you've added all of the cabbage to the container.

4. Cover the pressed cabbage with the plate and place the weight on top of the plate. This weight will force even more water out of the cabbage and will eventually keep the cabbage submerged under the resulting brine. Cover the container with the pillowcase or towel to keep dust and flies out.

5. Place the container in a convenient spot in your kitchen for a few hours.

6. After a few hours have elapsed, press down on the weight again to force a bit more water out of the cabbage. Continue doing this every few hours until the brine rises above the plate. This usually takes about a day or so. However, if you're using older cabbage, you might not extract enough water for this to happen. Don't panic. Mix about a tablespoon of salt per cup of water, stir until the salt is dissolved, and add enough of this mix to bring the brine level up above the plate.

7. Allow the sauerkraut to ferment for 2 days.

8. After 2 days, uncover your kraut container and take a look. Don't be surprised if you see bloom on the top of the brine. As long as the sauerkraut is below the brine, you have nothing to worry about. Skim the bloom the best as you can and rinse the plate and weight. Don't worry if you don't get every bit.

Now comes the question of how long you want to ferment your sauerkraut. It is basically a question of how tangy you prefer your kraut, so at this point you will need to start tasting it every day until it reaches your desired level of pucker. Just remember that you are limited to about two or three weeks in a warm environment. If you happen to have a cool cellar, placing your container there will buy you a few weeks' more time. Either way, if you try to go for the mouth-puckering level, you are courting disaster. If your kraut goes too long in a warm environment, it will eventually become soft and develop unpleasant flavors. When you're learning fermentation, it is best to go for a tasty tang without pushing the envelope, usually accomplished

Fermentation Factor

after about a week of fermenting. As you get more comfortable with the process, you'll get a feel for the right time to slow the process down. To slow the process, all you have to do is scoop out your sauerkraut, put it in clean containers, and store it in the fridge. As you store your kraut, make sure the brine level remains higher than that of the fermented cabbage and your homemade, healthy sauerkraut should keep for at least six months.

Variation: Russian Sauerkraut

For Russian sauerkraut, you can follow the above recipe, but add the following ingredients during step 2:
- 4 large carrots, peeled and grated
- 1 cup fresh cranberries
- 2 medium tart apples, diced
- 1 teaspoon sugar

Variation: Other Greens

You can also ferment other leafy vegetables besides cabbage. Fermented mustard greens are a popular Asian dish, and the process can be used for American mustard, collard, and turnip greens. Follow the same basic process for simple sauerkraut, except that there's no need to chop the greens. You can use the leaves whole, although some people prefer to rip the leaves into smaller pieces. Handle the leaves roughly as you pack them, as it will help release the juices to make the necessary brine.

Basic Kimchi

Kimchi is a fermented vegetable mix that originated in Korea, where it is still such a popular food that people eat it with almost every meal. Before starting, it should be noted that there are hundreds of variations of kimchi. You can make kimchi from fruits, roots, and a variety of vegetables; this is truly a food with a lot of space for creativity.

Photo courtesy of Wikimedia Commons by ayustety

Fermentation Factor

Equipment:
- Several medium-sized ceramic bowls
- Solid plate that will fit inside the bowls
- Clean quart-sized glass jar
- Good solid plate that will fit inside the jar
- Small weight that will fit inside the jar
- Cloth napkin, piece of cheesecloth, or small cloth towel

Ingredients:
- 4 tablespoons salt
- 4 cups water
- 1 pound bok choy
- 3 red radishes or 1 daikon radish
- 2 medium onions
- 2 medium carrots
- 3 tablespoons freshly grated ginger
- 4 cloves garlic
- 4 small red chili peppers or hot peppers of your choice

(Note: people who don't have hot peppers on hand might be tempted to use a tablespoon or two of that bottled hot sauce they've got in the fridge. Before giving in to that temptation, check the label of your sauce very carefully. If it contains chemical preservatives or words you can't pronounce, it will likely inhibit the fermentation process and ruin your kimchi.)

1. Mix the salt and water together in a bowl, stirring to thoroughly to dissolve the salt.

2. Chop the cabbage coarsely and slice the radish and carrots. Put these items in the bowl and pour in the salty water. Place the plate on top of the veggies to keep them submerged under the brine, and let the whole thing sit for 4 or 5 hours.

3. In another bowl, prepare the spices. Remove the seeds from the peppers and chop them into small pieces. Mince the onions and garlic cloves and grate the ginger. Crush these 4 items together to

Fermentation Factor

form a very tangy, pasty substance. (Wear gloves if peppers irritate your skin.)

4. Drain the brine off the softened vegetables. Set the brine aside in 1 of the bowls and taste the vegetables. You want them to be salty, but if they seem unpleasantly so, a quick rinse can bring the saltiness down a notch.

5. Mix the vegetables with the spice paste. Make sure the veggies are completely covered and then stuff this mixture into your quart jar. Pack it in good and tight until the brine rises to cover the mix. If the brine doesn't rise high enough, add a little of the brine you set aside earlier.

6. Cover the mix with a small plate and weigh it down. Cover the jar with a small towel and set aside.

Check on your kimchi daily by tasting a spoonful. Make sure you push the mix below the brine after you are done. After four or five days, it should achieve a pleasant spicy, sour flavor. At this point, all you have to do is place the jar in the fridge to slow down the fermentation.

Now that you understand the technique for making basic kimchi, let your imagination be your guide. Many root vegetables and fruits lend themselves to this versatile fermentation method. Experiment and enjoy!

Fermenting Beans

While beans are easy to store as dry goods, fermenting beans helps to unlock their nutrients and to create foods that are even more nutritious than rehydrated dry beans. Beans are not necessarily fermented to preserve them, but to make them more nutritious.

Photo courtesy of Wikimedia Commons by Brücke-Osteuropa

The fermentation process removes trypsin inhibitors that lurk on the coating of soybeans. These inhibitors interfere with the absorption of nutrients. In addition to this major benefit, traditional fermentation biologically enriches the food with protein, essential amino acids, essential fatty acids, vitamins, and numerous antioxidants.

You may be surprised to know that you already consume fermented beans on a regular basis.

Cocoa beans need to be fermented before they release the often-addictive flavor that we associate with fine chocolate. And without fermentation, vanilla beans are tasteless and odorless. High-quality vanilla beans are fermented for as long as six months. Although other processing methods are now being used in many countries, most coffee used to be fermented to remove the pulp that clings to the beans.

So, if you're a fan of chocolate, vanilla, or coffee, you've already had some exposure to fermented beans without venturing into the more exotic flavors of miso, natto, tempeh, fermented black beans, or any of the other wide variety of Asian fermented bean dishes.

Photo courtesy of WikiMedia Commons by David Monniaux

Most Asian bean ferments require a special starter like tempeh spore or koji. They also require a great deal of patience. A quality red miso can take at least two years to mature, and many gourmets prefer miso that has fermented for at least a dozen years.

The scope of this book doesn't allow time to delve into projects of that magnitude—there are plenty of books out there on the subject if you are interested. However, there are simpler bean ferments that will allow you to try your hand at this tasty dish. This savory bean paste is a tasty addition to breakfast burritos, salads, and tacos, and it plays together well with chips and salsa. You'll find a host of other uses for this easy-to-make fermented bean wonder.

Equipment:
- Wide-mouth quart mason jar
- Cheesecloth or a small towel
- Good strong rubber band
- Food processor, if desired

Fermentation Factor

Ingredients:
- About 3 cups of cooked and drained red, black, or pinto beans
- 1 red onion
- 5 peeled and crushed cloves of garlic
- 2 teaspoons salt
- 4 tablespoons whey (see chapter 6 for instructions on how to obtain whey)

There are two ways you can prepare this bean ferment. If you prefer a pasty texture, use a machine like a food processor to blend your ingredients. If you like a grittier consistency, mince and mash your ingredients by hand. It is simply a matter of preference; either way, your bean ferment will come out delicious.

1. Mince the onion and garlic (or chop them in a food processor).
2. Mash the beans and mix all the ingredients together, either by hand or with the food processor.
3. Place the mixture in the glass jar, making sure to leave at least 1 inch of air at the top.
4. Cover the jar with cheesecloth or towel secured with a rubber band.
5. Place the jar in a safe spot in your kitchen and let it ferment for 3 days.

The first bite might taste a bit fizzy, but that's fine. It's just an indication that it was fermenting. You can store any leftovers in the fridge for a week.

Pickled Okra

Equipment:
- Quart mason jar
- Piece of cheesecloth or towel
- Ingredients:
- 1 pound small okra (large okra can be tough, so choose the smallest you can find)

Photo courtesy of Wikimedia Commons by Gerard Cohen

Fermentation Factor

- 1 jalapeno, seeded and cut into small pieces
- 2 cloves garlic, peeled and cut into small pieces
- 1 tablespoon chopped fresh dill
- 1 tablespoon salt
- 1 cup water
- 4 tablespoons whey (see chapter 6 for instructions on how to obtain whey)

1. Wash the okra well and place in the mason jar.
2. Combine remaining ingredients and pour over the okra, adding more water if necessary to cover the okra. The top of the liquid should be at least 1 inch below the top of the jar.
3. Cover with the cheesecloth or towel and allow to ferment at room temperature for about 3 days. Move to cold storage after that. While your pickles will be ready to eat now, they will develop stronger flavors as you allow them to continue to ferment in the cold.

Green Bean Pickles

If you don't have a pressure canner to preserve green beans, this is a great option to store garden harvests for a little extra time.

Photo courtesy of Wikimedia Commons by FASTILY

Equipment:
- Large bowl
- Ceramic crock or other fermenting container with a capacity of a gallon or more
- Good solid plate that fits inside your container
- Your choice of weight
- Pillowcase or towel to use as a cloth cover

Ingredients:
- 2 pounds tender young green beans, washed and trimmed
- 6 garlic cloves, chopped
- 1 tablespoon pickling spice

- 6 dill heads
- Handful of dried peppers (optional, if extra spice is desired)
- ½ cup salt
- 3 quarts water

1. Mix together all ingredients except water and salt and add to the fermenting container.
2. Dissolve salt in the water and pour over the beans until they are covered.
3. Place the plate in the container, making sure it holds the beans down below the liquid. Use the weight to keep it in place.
4. Allow to ferment for approximately 2 weeks. If any bloom forms during that time, scoop it out and rinse the plate and weight. Refrigerate and enjoy!

Pickled Peppers

Equipment:
- Wide-mouth quart mason jar
- Smaller jar that can fit inside the mason jar
- Cheesecloth or small towel
- Mixing bowl

Ingredients:
- 4 cups fresh peppers (your choice)
- ½ onion, sliced
- 3 cloves garlic, roughly chopped
- 3 tablespoons salt
- 4 cups water

Before starting, make sure you've washed your peppers and that all are in good shape—remember that old or bruised produce will not make good pickles.

1. Combine the peppers, onions, and garlic and add them to the mason jar.

2. Mix together the water and salt and pour over the vegetables, ensuring that the brine completely covers them.

3. Place the small jar inside the larger one to hold the peppers below the brine. Add water to the smaller jar, if necessary, to keep it in place.

4. Cover with the cheesecloth or towel and allow to ferment for about 1 week before enjoying.

Pepper Sauce

Equipment:
- 2 quart-size mason jars
- Several pieces of cheesecloth
- Mixing bowl
- Food processor, if desired

Ingredients:
- 3 pounds fresh chili peppers (your choice) with stems removed
- 5 cloves garlic, peeled and minced
- 2 teaspoons salt
- 4 tablespoons whey

1. Combine and mince all ingredients until fine and pasty (use a food processor if desired).

2. Pour the paste into one of the jars and cover with a piece of cheesecloth. Ferment for 5 to 7 days.

3. After a week, place a large piece of cheesecloth in the mixing bowl and pour the fermented sauce in. Carefully pick up the cheesecloth and squeeze the sauce out, leaving any residue or pieces in the cloth.

4. Pour the strained sauce from the bowl into the clean jar and store it in the refrigerator. The sauce will keep for several months.

Fermentation Factor

Green Tomatoes

This is a great way to ensure that you "waste not" as your summer garden comes to a close. Right before the first frost, you can pick any remaining green cherry tomatoes and turn them into delicious pickled delights. Many gardeners get more cherry tomatoes than they could ever eat, so they pickle green tomatoes in the summer, to keep from being overrun with tomatoes.

Equipment:
- Wide-mouth quart mason jar
- Smaller mason jar (one that can fit inside the mouth of your other jar)
- Medium bowl
- Cheesecloth or small towel

Ingredients:
- 2 bay leaves
- 1 teaspoon pickling spice
- ¼ teaspoon celery seed
- 1 dill head
- 2 cloves garlic, roughly chopped
- ½ small onion, roughly chopped
- 2 cups green cherry tomatoes
- 4 cups water
- 3 tablespoons salt

Photo courtesy of Wikimedia Commons by Chris Haines

As with any fresh produce, make sure that you clean your green tomatoes, removing all stems, leaves, and dirt.

1. Poke each cherry tomato 2 times with a fork. This will allow the brine to thoroughly penetrate the skin.

2. Mix together the bay leaves, pickling spice, celery seed, dill, garlic, onion, and tomatoes and place in the quart mason jar.

3. Stir salt into water until dissolved and then pour over tomato mixture.

4. Use the smaller jar (filled with water, if necessary) to hold the tomatoes down below the top of the brine.

5. Cover with cheesecloth or small towel and allow to ferment for 5 to 7 days.

Like many other pickled products, these are ready to eat after the initial fermentation period; however, if you would like stronger flavors, you can place them in cold storage for up to a month before enjoying.

Fermented Citrus

Fermented citrus is popular in the Middle East and in Far Eastern cultures. Although lemons are the most popular to prepare, this recipe works with limes and tangerines as well. When the fruit is finished fermenting, all of it is edible (even the peels). Many enjoy this fruit for a snack, a condiment, or a tangy addition to a cocktail.

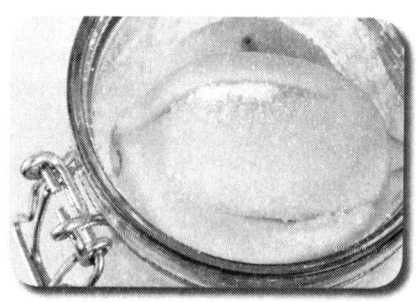

Equipment:

- Quart mason jar with lid
- Large wooden or plastic spoon
- Cheesecloth or small towel

Ingredients:

- 6 to 8 lemons, limes, or tangerines
- About ½ cup salt

In addition to these basic ingredients, you can also add spices, if you desire. Cinnamon sticks and peppercorns are popular additions.

1. Wash the fruit and make 2 long, intersecting cuts into each piece. The fruit should be cut as if you are quartering it, but it should still be attached at the top.

2. Sprinkle a bit of salt in the jar and about a tablespoon of salt inside each piece of fruit.

3. Place a layer of fruit in the jar and pound it with the spoon, releasing the juices. Sprinkle a bit more salt and add another layer of fruit, pounding it in again. Continue until the jar is full.

4. Add any spices that you desire and make sure the liquid completely covers the fruit. If not, pound the fruit a bit more to release the juice, or add a small amount of water to the jar.

5. Cover the jar with the cheesecloth and allow to ferment for 2 weeks. Every few days, cover the jar with the lid and shake well to distribute the salt and juice, then replace the cheesecloth. After the fermentation period is over, you can move your citrus to the fridge. Enjoy!

CHAPTER 4

Sweet and Sour
Fermented Breads and Grains

Breads

Photo courtesy of Wikimedia Commons by FASTILY

Although the tangy aroma and taste of sourdough bread often evokes thoughts of hardy miners seeking gold in the hills of California, the history of naturally leavened bread can be traced back to the ancient Egyptians. It's even mentioned in the Bible—when the Israelites left Egypt, they left in such a hurry that they didn't even have time to pack their starters. Hence, the eating of unleavened bread became a way of honoring the Passover.

Even today, some families treasure their sourdough starters as precious heirlooms, handed down from generation to generation. With proper care and feeding, a sourdough starter can remain viable for a very long time. The Boudin Bakery in San Francisco claims that it has been using the same sourdough starter since 1849!

While we can't claim that your sourdough starter will still be viable when the twenty-second century dawns, we are confident that a good starter will provide the basis for many loaves of healthy, hearty bread.

Using a naturally fermented sourdough starter has several advantages over using commercial yeast when it comes to making leavened baked goods, particularly since it is a self-sufficient process and does not rely on outside inputs for leavening. Since wild fermentation is a slower process, your dough has a chance to break down the gluten into nutrients that easier for your body to use. This process also adds beneficial B vitamins to the mix. While commercial yeasts stifle the growth of healthy microorganisms, naturally occurring yeast travels in the company of Lactobacilli and other acid-producing bacteria, which give your bread a wonderful symphony of flavors and aromas.

While there are many different ways to create a starter, the main goal is to create a growth of beneficial yeast that will both give rise to your bread and provide natural health benefits. In addition, this yeast will create conditions that encourage beneficial bacteria (like Lactobacilli) to grow while discouraging the growth of harmful ones.

Many sources recommend using whole wheat flour when creating a starter, as the natural yeasts found in the wheat will aid in the process. However, you can still be successful without whole wheat as long as you use unbleached flour with no additives for rising.

Sponge Starter

Equipment:
- Medium ceramic bowl
- Clean cloth that will cover the bowl
- Large spoon

Ingredients:
- About 2 cups unbleached all-purpose flour
- About ¼ cup water

1. Form about ¼ cup of flour into a mound in the bowl, making a small well in the center of the mound.

2. Place about 2 tablespoons of water in the well and slowly mix the flour into the water. You'll gradually be transforming this paste into dough.

3. Knead this small quantity of dough for about 5 to 8 minutes until it becomes springy.

4. Place the springy dough back in the bowl. Dampen your cloth and cover the bowl.

Fermentation Factor

5. Let the covered bowl sit in a warm spot in your kitchen for about two or three days.

6. Your starter is ready for the next step when it looks moist, wrinkly, and crusty. (This may be a bit confusing if you've never used a starter. Your starter will be a little dry and maybe even have a slight crusty feel to the outside. The interior of your starter will be moist, however.) When you pull off a piece of the crust, you should see tiny bubbles and smell a sweet aroma. If your starter meets those standards, it is ready to be refreshed. Pull any hardened crust off your starter. Add about ½ cup of flour and about 4 tablespoons of water to the bowl and mix this in with your starter.

7. Cover the bowl with a damp cloth again and set it aside for 1 to 2 days.

8. When you revisit the now expanded starter, pull off any sections of hardened crust and mix in another cup of flour.

9. Cover the bowl with a damp cloth again and set it aside for about 8 to 12 hours.

10. When the starter is complete, it will resemble risen dough. When you press it with a finger, it won't spring back.

11. Your starter is now ready to use. When you make your bread, make sure to pull off a small piece of the starter (about golf-ball size) and return it to the fridge. Next time you want to make bread, you'll already have the "seed." You just repeat steps 8-10, and you're ready to go.

Grape Starter

Grapes, along with many other types of fruits, contain natural yeast spores in the skin or peel of the fruit. You can use these natural yeasts to kick-start your bread starter.

Equipment:

- Quart glass or plastic container
- Several pieces of cheesecloth

Photo courtesy USDA

Fermentation Factor

Ingredients:
- Several unwashed grapes
- 3 cups unbleached all-purpose flour (plus more for additional feedings)
- 3 cups water (plus more for additional feedings)

1. Remove any stems or debris from the grapes, but don't wash them. This will remove the yeast that you need to start the fermenting process.
2. Crush the grapes by hand and place in a container covered with cheesecloth. Place in a warm spot out of the way and leave undisturbed for 3 days. After that time, you should start to see the liquid bubble, indicating that the yeast is growing.
3. Using a piece of cheesecloth, strain the solids out of the grape liquid and stir in 1 cup of flour and enough water to make a thick liquid.
4. Cover the container with a piece of clean cheesecloth and place in a warm spot for a day.
5. Discard all but 1 cup of the mixture and stir in 1 more cup of flour and 1 cup of water. Replace the cheesecloth and leave in a warm place for another day.
6. Repeat step 5 for another 1 to 2 more days. You should have a very bubbly starter at this point.

At this point, you can store your starter in the fridge or on the counter. If you leave it at room temperature, you will need to "feed" it about every other day. In the fridge, the starter will only need to be fed about once a week.

Variation: Apple Starter

The equipment and ingredients for making an apple starter are the same as the grape starter, except that instead of unwashed grapes, you will need a bit of grated apple. Mix the grated apple with one cup of water and one cup of flour; after letting it ferment for a day, continue with step 5 of the grape starter.

Potato Starter

This is perhaps one of the easiest starters to make, and it has the benefit of reusing ingredients from your kitchen that might otherwise go to waste. So next time you boil some potatoes, keep this starter recipe in mind.

Equipment:

- Quart glass or plastic container
- Cheesecloth or small towel

Ingredients:

- 1½ cups reserved water from boiling potatoes (allow to cool to room temperature)
- 1 tablespoon sugar
- 2½ cups unbleached all-purpose flour (plus more for additional feedings)
- Plain water for additional feedings

1. Mix together the potato water, sugar, and enough flour to make a thick liquid. Cover with the cheesecloth and set aside for a day.

2. Discard all but 1 cup of the mixture and stir in 1 cup of flour and 1 cup of plain water. Cover with the cheesecloth and allow to ferment for another day.

At this point you should have a healthy, bubbling starter. If it is still not bubbling well after the first two days, repeat step two for another day or two to allow the ferment to develop further.

Pineapple Starter

According to online sources, the idea for using pineapple juice in sourdough starter comes from chemist and baker Debra Wink (Breadtopia, http://www.breadtopia.com/make-your-own-sourdough-starter). After much experimentation with different types of liquids, she discovered that the acid level

Photo courtesy of Wikimedia Commons by Mitva Shah

in pineapple juice is perfect for encouraging the growth of wild yeast. If you find you are having trouble growing other starter cultures, you may want to give this one a try.

Equipment:
- Quart glass or plastic container
- Cheesecloth or small towel

Ingredients:
- 1½ cups unbleached all-purpose flour
- ½ cup unsweetened pineapple juice
- 1 cup water

1. Mix 2 tablespoons flour with 2 tablespoons pineapple juice in the container. Cover with the cheesecloth and let sit in a warm place for a day.
2. Stir in another 2 tablespoons of both flour and pineapple juice to the mixture. Cover with the cheesecloth and return to the warm place for another day. You may begin to see bubbling at this point, but don't give up if you don't.
3. Stir in yet another 2 tablespoons of both flour and pineapple juice to the mixture. Cover again and return to a warm place for a third day.
4. On the fourth day, stir the mixture well and measure out ¼ cup, discarding the rest. To the reserved starter, add ¼ cup of flour and 2 tablespoons of water. Cover and return to a warm place for another day.
5. Repeat step 5 until the mixture double in size and starts to smell yeasty.

If your mixture starts to bubble, then suddenly quits, some bakers recommend adding ¼ teaspoon of cider vinegar with one of the daily "feedings." In some cases, this can reactivate your starter.

Once the yeast starts growing, your starter should be fed equal parts of flour and water (about the same amount of each as the amount of starter present—so, for example, if you have one cup of starter, you will want to add one cup of flour and one cup of water). Store the starter in the refrigerator when you are not using it. It needs to be fed once a week to keep it alive. Feel free to use or discard some starter when needed, otherwise you will be overwhelmed with starter very quickly!

Your starter will be ready to use once it is actively bubbling; however, it will "mature" over several weeks and will provide better rising properties then.

Drying Yeast for Later

Of course, there are times when having a liquid starter sitting around just isn't convenient. You don't find too many jars of liquid starter at your local grocery store—you find dried yeast. This is because it will last a lot longer and can go without "feeding" when it is dried, making it a lot easier to keep around. Luckily, you can make your own dried yeast at home too.

Take any of your starters, line a cookie sheet or baking stone with waxed paper, and spread the starter in a thin layer. (Although this works best with the liquid-style starters, you can also use a sponge starter—just make sure you take care to spread it as thin as you can.) You then have several options for dehydrating it. First of all, if you have a dehydrator, feel free to use it. Just spread the starter on a piece of waxed paper in the trays of your machine instead of on a cookie sheet or baking stone. If you live in a hot and dry climate, you may just be able to cover it with a piece of cheesecloth and place in the sun. Otherwise, set your oven to the lowest temperature and heat until just dry. Make sure you don't overheat or cook the starter, as this will kill the yeast and make it completely useless. Once the yeast is dry, you can crumble it and store it in an airtight container. Just like store-bought yeast, it will last longer in the fridge or freezer.

Play around with amounts you use in recipes once the yeast is ready, as the potency of homemade yeast will be a little different than the store-bought version. You will probably need more of it for the same amount of bread—try doubling it to start, then adjust from there.

Sourdough Bread

There are almost as many recipes for sourdough bread as there are people who love it. Included here are several different types of sourdough bread, including loaves, baguettes, rolls, biscuits, and even pancakes. The options for sourdough bread are endless, so if something here doesn't suit your fancy, keep looking. There are many resources available to you, including cookbooks and online recipe sites.

San Francisco-Style Sourdough Bread Recipe

Equipment:
- Large mixing bowl
- Plastic wrap
- Pizza peel or cookie sheet with no raised edges
- Cookie sheet, baking stone, or two glass or ceramic bowls
- Small bowl or saucepan
- Small metal pan
- Pastry brush
- Cooling rack

Ingredients:
- 2½ cups warm water, plus additional water during baking
- 1 cup sourdough starter
- 6 to 8 cups flour (if you use a sponge starter, you will need less than if you use a liquid starter)
- 2 teaspoons salt
- 2 teaspoons sugar
- ½ teaspoon cornstarch

1. In a large bowl, combine 2 cups of water, the sourdough starter, and 4 cups of flour. Mix well and cover with plastic wrap in a warm place for 8 to 12 hours (overnight is best).

2. Mix in the salt and sugar. Add flour a little at a time to make a very stiff dough and knead until smooth. You may not use all the flour. Cover the dough and let it rise for 2 to 3 hours.

3. Punch down the dough and divide in half. Knead each piece until smooth and then form it into a round.

4. If you are planning to bake your bread on baking stone, dust your pizza peel or edgeless cookie sheet well with flour and place the loaves onto it. If you are using a cookie sheet or ceramic bowl to

bake your bread, grease the container well and place the bread on it. Cover lightly and let the bread rise until doubled in size and puffy (about 1 to 2 hours).

5. While this is rising, mix the remaining ½ cup of water and ½ teaspoon cornstarch in a small bowl and heat until boiling. Remove from heat and let cool.

6. 30 minutes before time to bake, heat your oven to 400°F. If you are using a baking stone, you will want to place the stone in the oven now to allow it to heat with the rest of the oven.

7. Carefully place a pan with about 2 cups of water on the bottom rack of the oven (this will help make the crust of your bread nice and crispy). Cut several slashes across the top of each loaf, and place in the oven. (If you are using a baking stone, slide the loaves off the floured cookie sheet and onto the hot stone.) Bake for 10 minutes.

8. Pull out the rack and brush the tops of each loaf well with the cornstarch mixture. Close the oven and allow to bake for another 20 or 25 minutes. Loaves should be a light golden color and sound hollow when tapped on the bottom. Remove from pan and allow to cool completely on a cooling rack before slicing.

French-Style Sourdough Bread

Equipment:
- Large mixing bowl
- Stand mixer or heavy spoon
- Clean towel or cloth
- Plastic wrap or shower cap
- Baking stone or cookie sheet
- Pizza peel or edgeless cookie sheet (if using a baking stone)
- Sharp knife
- Small metal pan
- Cooling rack

Photo courtesy Wikimedia Commons by David Monniaux

Fermentation Factor

Ingredients:
- 1 cup sourdough starter
- 4½ to 5 cups unbleached all-purpose flour
- 1¾ cups water, plus additional water during baking
- 2 teaspoons kosher salt

1. In a large mixing bowl, combine the flour, water, and starter until everything is mixed together. Let the mixture rest for about 30 minutes.

2. Add in the salt and knead for five to seven minutes. Cover with a clean towel or cloth and let the dough rest at room temperature for 1 hour.

3. After an hour, knead the dough for 3 to 5 minutes and then cover with the towel again. Let rest for 1 hour and then repeat this short kneading process again. Let rest 1 more hour.

4. Remove the towel over the dough and replace with the plastic wrap or shower cap. Place in the fridge at least overnight (24 hours is best). The dough should be doubled in size by the time you remove it.

5. Remove the plastic wrap or shower cap from the dough and recover the dough with the towel. Allow the dough to return to room temperature.

6. After about an hour, preheat the oven to 450 degrees and place the dough on a well-floured surface. Divide the dough into 4 equal pieces using a sharp knife. Shape each piece into a log (or baguette), making sure it is still small enough to fit on your baking stone or cookie sheet. If you will be using a baking stone, you should dust the pizza peel or edgeless cookie sheet with flour and place the loaves there. If you will be using a cookie sheet, flour that cookie sheet and place the loaves on the sheet. Allow them to rest for 20 to 30 minutes.

7. Pour 1 to 2 cups of water in the small pan and place on the lowest rack of your oven. Use the sharp knife to put several slashes on each loaf. If you are using the baking stone, slide the baguettes off the pizza peel or edgeless cookie sheet and onto the baking stone.

If you are using a cookie sheet to bake the bread, simply place the sheet in the oven.

8. Bake the loaves for 20 to 25 minutes, or until golden brown. Loaves should sound hollow when tapped on the bottom. If the loaves don't brown evenly, rearrange the loaves in the oven after about 15 minutes of baking. Allow to cool completely on a cooling rack before enjoying.

Simple Sourdough Rolls

Equipment:
- Mixing bowl
- Cookie sheet or muffin tin

Ingredients:
- 1 cup self-rising flour (if you only have all-purpose flour, add ¾ teaspoon of baking soda,
- 2 teaspoons baking powder, and ½ teaspoon salt)
- 1 cup sourdough starter
- ½ cup vegetable oil

Combine all ingredients and mix well. Shape into 1½ inch balls. Bake in a greased muffin tin at 350 degrees for 25 to 35 minutes or on a greased cookie sheet at 425 degrees for 10 to 15 minutes.

Sourdough Biscuits

Equipment:
- Mixing bowl
- Cookie sheet

Ingredients:
- 1 cup sourdough starter
- 1 cup self-rising flour (if you only have all-purpose flour, add ¾ teaspoon baking soda and ¼ teaspoon salt)
- 1/3 cup softened margarine or cooking oil

Mix all ingredients together and drop by spoonfuls onto a greased cookie sheet. Bake at 350 degrees for 10 to 20 minutes.

Sourdough Pancakes

Equipment:
- Mixing bowl
- Large skillet (preferably non-stick)
- Spatula

Ingredients:
- 1 cup sourdough starter
- 1 cup self-rising flour (if you only have all-purpose flour, add ¾ teaspoon baking soda and ¼ teaspoon salt)
- 1/3 cup softened margarine or cooking oil
- 1 egg
- Enough milk for desired consistency

Mix all ingredients together and pour batter by small amounts into a greased skillet heated to medium-high heat. When bubbles form and burst in the batter, use a spatula to check the bottom of the pancake for doneness. If the pancake is sufficiently browned, flip the pancake over quickly and cook until the other side is browned. Butter immediately and serve with your toppings of choice. Don't forget to re-grease the skillet periodically as you continue cooking your pancakes.

Amish Friendship Bread

Amish friendship bread is very likely not even Amish, but it is a great treat to share with your friends. The recipe produces a sweet bread, much like muffins or a cake. The best part is that there are dozens of ways to modify the recipe, so it's different every time you make it. Don't be intimidated by the long recipe. Although feeding the starter takes a little time, actually making the bread is quite easy.

The Starter

If you already have starter from a previous batch or from a friend who has shared with you, skip this step.

Fermentation Factor

Equipment:
- Quart mason jar or gallon re-sealable bag
- Cheesecloth (only if using mason jar)
- Ingredients:
- ½ cup liquid bread starter
- ½ cup milk
- ½ cup sugar
- ½ cup unbleached all-purpose flour

Combine all your ingredients inside the container and mix well (if using a re-sealable bag, mush the bag). Mix as best as you can, but don't worry if there are a few lumps. The yeast will munch those up in no time. You are now on day one!

The Feeding Process

As you feed and grow your Amish friendship bread, you need to make sure it stays somewhere warm. Resist the temptation to put it in the fridge.

Equipment:
- Quart mason jar or gallon re-sealable bag
- Cheesecloth (only if using mason jar)
- Wooden or plastic spoon
- Small plate or tray (only if using mason jar)

Ingredients:
- 1 cup Amish friendship bread starter
- 1 cup unbleached all-purpose flour
- 1 cup sugar
- 1 cup milk

The day you created your starter (or put some aside after bread making) is considered day one. Place the starter in a quart mason jar covered with cheesecloth or a gallon re-sealable bag. It is a good idea to write down the date you do this, as it is easy to lose track of the feeding day you are on otherwise.

Fermentation Factor

1. On day 2, use a wooden or plastic spoon to stir your starter (or mush the bag). If you are using a re-sealable bag to store your starter, you will also want to "burp" your bag at this time. Place your starter back in its warm, safe place. Repeat this step on days 3, 4, and 5.

2. On day 6, mix in the flour, sugar, and milk. Don't worry if there are any lumps left after you've mixed it; the yeast will eat those quickly enough. If your starter is in a mason jar, you may want to place it on a small plate or tray—occasionally the fermentation process will become active enough that the starter may overflow a bit.

3. On day 7, use a wooden or plastic spoon to stir your starter (or mush the bag). Don't forget to "burp" your starter if you are using a re-sealable bag. Repeat this step on days 8 and 9 as well.

Baking the Bread

You have two options at this point. Part of what makes this treat "friendship bread" is the fact that you share it with friends. If you have a few friends who might want some starter to make their own Amish friendship, go with the first option. If all of your friends have already been inundated with bread starter and you don't need any extra, then choose option 2.

*Option 1: Bread plus starter to share

Equipment:
- Glass or plastic bowl
- Wooden or plastic spoon
- Glass or plastic measuring cup
- Three 1 gallon re-sealable bags
- Quart mason jar or gallon re-sealable bag
- Cheesecloth (only if using mason jar)

Ingredients:
- Amish friendship bread starter (fed and tended for ten days)
- 1½ cups unbleached all-purpose flour
- 1½ cups sugar
- 1½ cups milk

Fermentation Factor

Pour the starter into the bowl and add the flour, sugar and milk. Stir well. Measure 1 cup of the starter and place it into each of the three 1 gallon re-sealable bags. These are to give away to friends. (Don't forget to include the recipe!) Place one more cup of starter in the quart mason jar or gallon re-sealable bag and set aside. This is your starter to keep for more batches of bread, which is now back on day 1. Use the remaining starter for the recipe below.

Option 2: Bread with no extra starter

Equipment:
- Glass or plastic bowl
- Wooden or plastic spoon
- Glass or plastic measuring cup
- Quart mason jar or gallon re-sealable bag
- Cheesecloth (only if using mason jar)

Ingredients:
- Amish friendship bread starter (fed and tended for 10 days)
- ½ cup unbleached all-purpose flour
- ½ cup sugar
- ½ cup milk

Pour 1½ cups starter into the bowl and discard the rest. Add the flour, sugar, and milk to the bowl and still well. Pour 1 cup of the starter into the quart mason jar or gallon re-sealable bag. This is so that you can continue to make more bread, so consider the starter back on day 1. Use the remaining starter in the bowl for the recipe below.

The Recipe

One of the great things about this recipe is the variety it allows. Start with the basic recipe the first time you make the bread (just to get the hang of things). After that, be creative. Instead of vanilla pudding, try banana pudding and mix in a cup of chopped nuts to make banana nut friendship bread. You could also leave out the cinnamon and use lemon pudding for a twist. Many people like to add an extra two teaspoons of cinnamon and a cup of raisins for a delicious cinnamon-raisin bread. Choose your favorite flavors and have fun!

Fermentation Factor

Equipment:
- 2 loaf pans
- Mixing bowl

Ingredients:
- Remaining starter from option 1 or 2
- ½ cup oil
- ½ cup applesauce
- ½ cup milk
- 3 eggs
- 1 teaspoon vanilla
- 2 cups self-rising flour (if all you have is all-purpose flour, simply add 1½ teaspoons of baking powder, ½ teaspoon of salt, and ½ teaspoon of baking soda)
- 1½ cups sugar
- 2½ teaspoons of cinnamon
- 1 large box of vanilla instant pudding mix

1. Take the starter you have in your bowl from option 1 or 2 and add the oil, applesauce, milk, eggs, and vanilla. Stir well.

2. In a separate bowl, mix together the flour, pudding mix, 1 cup of sugar, and 2 teaspoons of cinnamon. Note that you are not using all of the cinnamon and sugar at this point. Add this dry mixture to the wet mixture.

3. Grease the 2 loaf pans. Mix the remaining ½ cup of sugar and ½ teaspoon of cinnamon together. Cover the bottom of the loaf pans with this sugar mixture, and then pour in the batter and sprinkle the remaining mixture on top.

4. Bake at 325 degrees for 1 hour or until done. A knife inserted in the middle of the loaf should come out clean when the bread is done. Loaves freeze well (if they last that long).

Fermented Oat Porridge

Equipment:
- Small bowl
- Cheesecloth or small towel
- Small boiler
- Spoon

Ingredients:
- ½ cup rolled oats
- 3½ cups water
- Pinch of salt
- Your choice of flavorings (fruit, cinnamon, honey, etc.)

1. Cover the oats with 1 cup of water in the small bowl and cover with the cheesecloth. Allow to ferment for about 24 hours.
2. Boil the remaining water and add the salt. Once boiling, add the fermented oats.
3. Reduce heat to simmering, stirring often for 10 minutes or until thickened.
4. Remove from heat and add your choice of flavorings. Enjoy!

Fermented Granola

Equipment:
- Large glass or plastic bowl
- Medium mixing bowl
- Sturdy wooden or plastic spoon
- Cheesecloth or towel
- Dehydrator (optional)
- Cookie sheets
- Wax paper

Ingredients:
- 4 cups rolled oats

Fermentation Factor

- 2 cups raisins
- 1 cup chopped almonds (or your nut of choice)
- 1½ teaspoons salt
- 1 teaspoon cinnamon
- ¼ teaspoon nutmeg
- 1 cup hot water
- 1¼ cups yogurt (see chapter 6 for instructions on how to make your own yogurt)
- ¼ cup honey
- ¼ cup butter, softened
- ¼ cup coconut oil

1. In the large bowl, combine the oats, raisins, nuts, salt, cinnamon, and nutmeg.
2. In the medium-sized mixing bowl, combine remaining wet ingredients.
3. Using the wooden spoon, fold the wet ingredients into the dry ones, stirring until just moistened.
4. Cover with cheesecloth or towel and allow to ferment for 2 days.

You have two choices for your dehydration. You can either use a dehydrator or your oven—whichever you prefer. If you are using a dehydrator, follow the instructions below, but instead of placing your batter on the cookie sheets, place it on the trays of your dehydrator.

5. Line the cookie sheets with wax paper to prevent sticking.
6. Handling the wet granola batter gently, drop ½ inch spoonfuls of granola onto the cookie sheets, leaving plenty of space between lumps for air to flow.
7. Bake at 145 degrees for 1 hour. If your oven will not go this low, set it at its lowest temperature and check your granola more frequently. You can also leave the door of your oven slightly cracked to reduce the temperature inside while baking.
8. Break granola lumps in half to allow air to reach the centers better. Reduce temperature to 125 degrees and bake for 1 more hour.

Fermentation Factor

(Again, if your oven will not go this low, set it as low as it will go and just check your granola more frequently.)

9. Crumble your granola into small pieces (as chunky or fine as you like). Continue to bake your granola until it is completely dry.

CHAPTER 5

Delicious Drinks
Fermented Beverages

When most people think of fermenting, their mind immediately jumps to fermented beverages like beer and wine. There are almost innumerable varieties of alcoholic drinks that you can create. The process for creating each drink is unique, and dozens of factors go into each batch. That is why a vineyard can produce wine the same way each year but still have vintages that are better than the others.

Most people become intimidated at the thought of fermenting their own drinks, mainly because it seems like there are so many things you have to do and buy. Isn't there special equipment you need? Special yeasts? Particular grains or fruit grown only in remote locations?

Yes and no. Yes, many brewers and winemakers utilize all these things, and they contribute to finer products, but no, they aren't required. The fact that more than one prisoner has been caught making booze in their cell with little more than fruit cups, orange juice, and trash bags proves that you can make do with almost anything and still get alcohol.

While the scope of this book doesn't really provide the opportunity to delve into brewing beer or aging wine, there are plenty of resources out there than can assist you in these pursuits. In the mean time, we will include a couple of simple alcoholic beverages you can try—t'ej and elderflower wine.

T'ej

T'ej is a honey wine (or mead) from Ethiopia. Traditionally, it is made with a plant called gesho, which is somewhat like hops. Although it is available in the United States, it's probably not on the shelf of your local grocery store (unless your local grocery store is an Ethiopian market). Therefore, this recipe (which doesn't include the gesho), is not true t'ej in the strictest sense. However, it gives you a good foundation for experimenting with brewing your own alcoholic beverages.

The only ingredients required for t'ej are water and honey, but you can create your own unique twist by adding additional spices and fruit. Be creative! Some have made t'ej with bananas, peaches, apples, even coffee, along with spices like cinnamon and ginger. There are other variations in the previously mentioned Wild Fermentation. It seems your only limit is your imagination in this regard.

Equipment:
- Boiler (large enough to hold a gallon of liquid)
- Wide-mouthed glass container (at least one gallon)
- Slotted spoon
- Glass jug (at least one gallon)
- Airlock or balloon

Ingredients:
- 3 cups raw honey
- 12 cups water
- Unwashed organic berries or grapes (optional)
- 1 cup fruit, your choice (optional)
- Your choice of spices (optional)

Fermentation Factor

To make your own batch of t'ej:

1. In a large boiler, heat the water to boiling and add spices. Remove from heat, cover, and allow to steep for 10 minutes. Stir in honey and cool to room temperature. Strain out spices.

2. Place berries/grapes and fruit in the bottom of a wide-mouth glass container (at least 1 gallon). Add strained honey mixture and cover with cheesecloth. Set it a warm place and allow to ferment for 2 days, stirring 2 times each day.

3. After 2 days, scoop out the fruit with a slotted spoon. After 3 or 4 days, the brew should start to become bubbly and can be transferred to a jug. Technically it is drinkable at this point, although the bubbling will make it rather fizzy (and it will not have had much time to develop much of an alcohol content). If you'd rather let it continue fermenting and aging, it will continue to bubble and develop flavors for several more weeks.

A note: Most brewers use a device called an airlock on jugs to allow gasses out of the bottle without letting additional air in. Capping a bottle that is still fermenting will result in disastrous consequences. Airlocks can be purchased fairly cheaply, or you can use a balloon in a pinch. Simply place the neck of the balloon over the opening of the jug. Make sure to "burp" it every few days if you choose to use this method!

Elderflower Wine

Equipment:

- 2 clean food-grade buckets
- Clean towel or cloth
- Cheesecloth or sieve for straining
- Bottles for storing

Ingredients:

- 1 gallon hot water
- 1½ pounds (about 3½ cups) sugar
- ½ gallon cold water
- Juice from four lemons

Fermentation Factor

- Zest from four lemons
- 2 tablespoons white wine vinegar
- 15 to 20 elderflower heads (elder flower comes from the elder shrub—it is an herb)

1. Pour the hot water into the bucket and stir in the sugar until completely dissolved.
2. Mix in the cold water, lemon juice, lemon zest, and white wine vinegar in a separate bucket. Stir the lemon-vinegar mix into the bucket of sugar water and add the elderflower heads.
3. Cover the bucket with a clean tea towel and leave in a cool place to ferment. Check on the bucket every few days. When it starts getting foamy you know that it is fermenting and you should leave it for another 4 or 5 days.
4. Strain your wine and decant it into sterilized bottles for storage. Store the wine in a cool place for at least 1 week before drinking.

Other Drinks

There are other fermented beverages that are not alcoholic and provide many health benefits. Kombucha is a fermented tea that has seen a resurgence in popularity. While relatively easy to make, kombucha requires a "mother," which is a piece of starter used to begin the fermentation process. Unfortunately, a mother can be difficult to locate. There are other options that offer similar benefits and are easier to make.

Rejuvelac

Dr. Ann Wigmore founded the Hippocrates Health Institute in Boston, Massachusetts in 1958 and became one of the leading advocates of living foods for living bodies. While working with sprouted wheat berries, she noticed that the water from a recently soaked batch had a pleasant smell. Further experimentation led her to create a sprouted grain beverage that contains gut-enhancing probiotics and numerous nutrients, which she named rejuvelac. Some people claim that drinking three glasses of rejuvelac a day helps promote weight loss. Whether this is true or not, this refreshing, energizing, simple-to-make drink is a healthy addition to any diet.

Fermentation Factor

Equipment:
- A wide-mouth two-quart glass jar
- A small section of cheesecloth (enough to cover the jar and secure with a rubber band)
- A sturdy rubber band
- A quart-size glass measuring cup

Ingredients:
- 6 to 8 quarts water
- 1 cup wheat berries

1. Rinse the wheat berries to remove any dust and debris.

2. Place the wheat grain in the jar and add enough water to almost fill the jar. Cover the jar with cheesecloth and secure with the rubber band. Let the berries soak for about 12 hours, then strain the water through the cheesecloth and discard.

3. With the cheesecloth securely attached, set the jar upside down in the measuring cup. Make sure it rests high enough above the bottom of the cup to avoid exposing the wheat berries to any water that drains into the cup. If you drained the berries well, this shouldn't be a problem.

4. Rinse the berries at least 2 times a day in water. You don't want them to dry out, so you might need to rinse them more often if your room is warm. After about 2 or 3 days, you should see little tails emerging from the wheat berries, meaning you have created your sprouts.

5. Rinse the sprouted berries with water, return them to the jar, and cover them with about 2 quarts of water. Cover the jar with the cheesecloth again and let it ferment for 2 days.

6. Pour the liquid off the sprouted berries and store in a separate glass container of your choice. This is your rejuvelac, and it will keep in your fridge for several days. Good rejuvelac is a cloudy, slightly yellow liquid with a tart, lemony flavor.

7. If you want to continue making more, return to the jar with the sprouted wheat berries and add another 2 quarts of water. Re-cover with the cheesecloth and set aside for about 1 day, after which you'll have another 2 quarts of rejuvelac.

Some people claim that you can repeat this procedure to make a third batch, but it's not recommended. Sprouting another cup of wheat berries will lead to better results. In addition to wheat berries, rejuvelac can also be made from barley, oats, rye, unhulled millet, brown rice, and raw unhulled buckwheat. Feel free to experiment to find the flavor you like best.

Root Beer

Fermented soft drinks are a tasty delight and a great way to introduce your kids to the great world of fermenting. However, there is one question that may be burning in your mind right now: If fermenting sugary liquids can make alcoholic drinks like wine and beer, does that mean that these soft drinks are alcoholic?

Technically, yes—but before you start reading excitedly or angrily flipping to the next chapter (depending on your thoughts on the matter), it should be noted that the alcohol content in these drinks is extremely minute. You or your kids could probably drink the whole two-liter bottle and never notice a thing. However, we do want to mention that there are traces of alcohol that remain from the fermenting process, just in case even this small amount is problematic due to medical or religious reasons.

Another quick note: homemade soft drinks can be extremely fizzy when opened for the first time. Open the bottles slowly and with care if you don't want a mess.

With that being said, there are many ways to make homemade root beer, including using your own roots. However, some of these flavorings can be a bit difficult to find, depending on where you live, so this recipe will use root beer extract instead. If you are interested in making root beer the truly old-fashioned way, there are plenty of online resources to get you started.

Fermentation Factor

Equipment:
- Clean 2 liter soda bottle with a screw cap
- Funnel

Ingredients:
- 1 cup sugar (you can use more or less, depending on your tastes)
- ¼ teaspoon powdered yeast (feel free to use your own homemade dry yeast from the last chapter, although remember that you may need a little more of that to get the job done)
- 1 tablespoon root beer extract (again, use more or less depending on how strong you like your root beer)
- Water to fill the bottle

1. Using the funnel, pour in all the ingredients except the water.
2. Fill the bottle about half full with water. Cap the bottle and swirl or shake it until the contents are dissolved.
3. Finish filling the bottle with water and recap it tightly. Let the bottle sit at room temperature for about four days or until it feels hard (when you press on the bottle, the sides shouldn't give). Once this happens, move it to cold storage immediately. Once it's cold, enjoy your homemade root beer!

Variation: Cream Soda

Follow the directions for making root beer, substituting one tablespoon of vanilla extract for the root beer extract. Depending on your tastes, you may want to adjust this amount.

Variation: Ginger Ale

Follow the directions for making root beer, except substitute one tablespoon of ginger ale extract for the root beer extract. If you can't find ginger ale extract, you can also use 1½ tablespoons of freshly grated ginger. Feel free to adjust the amounts to match your tastes.

Vinegar

Making vinegar is as simple as picking up a bottle of preservative-free apple cider, removing the cap from the bottle, and covering the opening with a piece of cheesecloth secured with a rubber band.

If you let that bottle sit on your counter for three or four weeks, you'll end up with a bottle of apple cider vinegar.

The ancient Greek physician Hippocrates was among the first to promote apple cider vinegar as a powerful cleansing and healing elixir. Various studies since then have indicated that apple cider vinegar is effective at preventing arthritis, osteoporosis, and infections. It has also been used to soothe burns (including sunburns), soothe itches, aid digestion, preserve memory, and control weight. It can even be used to treat hives and other allergic reactions. Pretty potent stuff!

Making vinegar is a very effective way to make something useful from bruised fruit, overripe bananas, apple peels, apple cores, wrinkled blueberries, and lots of other scraps that might otherwise be headed to the trash or the compost heap. It only takes a bit of time and patience to enjoy the complex flavor of your very own homemade vinegar.

Equipment:
- Medium-size ceramic bowl or large glass jar
- Cheesecloth to cover the container
- A good strong rubber band to secure the cheesecloth
- A sturdy plate that will fit in the container
- A small jar that will fit inside the bowl or jar

Ingredients:
- About 1 quart water
- ¼ cup sugar or honey
- 2 cups fruit peels or bruised fruit, coarsely chopped

1. Pour the sugar or honey into the bowl or jar. Add the quart of water and stir until it is all dissolved, then add the fruit.

2. Place the plate on top of the fruit mixture and fill the small jar with water so that it serves as a weight. Place it on top of the plate,

making sure that it has all the fruit pushed down below the water line.

3. Cover the container with the cheesecloth and secure it with the rubber band. Let this mixture ferment for about a week.

4. When the liquid turns dark, strain the fruit matter out of the liquid and discard it. Let the liquid ferment for another three weeks. Stir the liquid every 2 or 3 days.

5. Pour the new vinegar into a clean glass jar or bottle to store.

Unlike many sterile-tasting commercial vinegars, this homemade vinegar will have a rich variety of flavors that will enhance any dish you use it in.

CHAPTER 6

Curds and Whey
Fermenting Dairy

Simple Cheese

Some people dedicate their entire lives to the fine art of making cheeses. This popular fermented dairy product comes in so many different varieties that just learning the names and qualities of them all can be an intense course of study.

Although mastering all the methods and means of making exotic cheeses requires investing in the proper equipment and acquiring a spectrum of specialized bacteria, there are some forms of cheese that can be created with common ingredients and items you probably already have in your home.

One favorite is a soft and simple farmer's cheese that goes well with good bread, fresh fruit, or as a filling in baked goods. Best of all, it is easy to make, and you can customize it with your flavor touches ranging from chives to ground peppercorns.

Equipment:
- Large cooking pot
- Wooden spoon
- Large colander or sieve
- 2 ceramic or glass bowls
- A yard or two of cheesecloth
- Airtight container
- Candy thermometer (if desired)

Ingredients:
- 1 quart of whole milk

Fermentation Factor

- 1 cup of buttermilk
- 2 or 3 teaspoons of lemon juice
- ¾ teaspoon of salt

1. Pour the whole milk into the large pot and heat at a medium-low setting until you see small bubbles forming around the edges (about 175 degrees if you are using a thermometer). Don't forget to stir the milk occasionally while it is heating to avoid it burning on the bottom of the pot.

2. Turn off the heat and add the buttermilk and lemon juice. The milk should begin to curdle. If your milk mix doesn't begin curdling within 30 seconds, add 1 more teaspoon of lemon juice.

3. Line the colander with a double layer of cheesecloth. Place the colander or sieve over a bowl and pour the curdled milk through the cheesecloth to catch the curds. You have now produced curds (in the cheesecloth) and whey (in the bowl).

4. Let the colander drain for about 10 minutes and then give the cheesecloth package a light squeeze to remove more liquid. Drain for another 20 to 30 minutes.

5. Place the cheese in the bowl. Break it up and mix in the salt. At this point, you can also add any flavors you choose.

6. Store your farmer's cheese in the airtight container and place it in the fridge for a few hours. The time it spends in the fridge is just as important as the rest of the process, as this is the time when the bacteria that give the cheese its flavor get a chance to begin growing.

You should eat this tasty cheese within a week of creating it, but you'll probably find so many uses for it that it won't even last that long.

Labneh

Labneh is a traditional Middle Eastern cheese made from plain yogurt and stored in olive oil. You can use store-bought yogurt if you like, or you can use the yogurt you make at home (see later in this chapter).

Equipment:

- Mixing bowl
- Cheesecloth
- Colander or sieve
- Rubber band or piece of kitchen twine
- Bowl (big enough to rest your colander on)
- Small mason jar

Ingredients:

- 1 quart plain yogurt
- ½ teaspoon kosher salt
- Olive oil and herbs, if desired
- Mix the yogurt and salt together in the mixing bowl.

1. Set the colander or sieve on top of the bowl and line it with the cheesecloth. Carefully pour in the yogurt mixture into the cheesecloth. Allow to drain for about 10 minutes.

2. Carefully gather the edges of the cheesecloth together and gently twist them, forming a ball in the center. Tie the edges with the rubber band or string.

3. Return the cheesecloth to the colander and allow to drain for at least 12 more hours (24 is better). Make sure the dripping whey level never makes it back up to the top and touches the colander. It will not drain properly if it does. (Remember to store this whey in the fridge for later use!)

4. After your labneh has drained, carefully unwrap the cheesecloth. You can store the labneh alone in a mason jar or in olive oil with herbs. If you choose to store it in olive oil, carefully roll the labneh

into small, walnut-size balls, place them in a mason jar with the herbs of your choice, and cover with the oil.

Ricotta

Ricotta is Italian for "cooked twice" or "cooked again" and is made by heating leftover whey. When you strain the liquid off the labneh, there are still plenty of cheese-making proteins left in the whey that can be used, and they make a delicious soft cheese that is great for cheesecakes or lasagna. It is also much tastier and healthier than anything you can buy in a plastic tub from your local grocery store.

A note: You don't want to use the whey that you get from the farmer's cheese at the beginning of the chapter to make ricotta, as the acid in the lemon juice has already stripped most of the proteins out of that whey during the cheese-making process. Use the whey from the labneh instead.

Photo courtesty Wikimedia Commons by Paoletta S.

Equipment:
- Large pot
- Candy thermometer
- Colander or sieve
- Large bowl that can hold the colander
- Fine cheesecloth or other cloth
- Rubber band or kitchen twine
- Mixing bowl

Ingredients:
- 2 quarts whey
- 1 tablespoon vinegar
- Salt to taste

1. Pour the whey into the pot and heat to 200 degrees. (If it reaches boiling, the world won't come to an end, but try not to let it happen.)

2. Add in the vinegar and stir gently. You should begin to see small white particles floating in the whey as the grains of ricotta begin to form.

3. Line the colander or sieve with the cheesecloth and place it in the bowl. Carefully pour the hot liquid into the cheesecloth, then gather up the edges and tie them together. Return to the colander.

4. Allow the ricotta to drain for several hours, making sure that the draining whey does not reach the colander or cheesecloth. (The longer you let the ricotta drain, the drier it will be.)

5. Once your ricotta is finished draining, place it in the mixing bowl, break apart the curds, and add salt to taste. Your fresh ricotta cheese will stay fresh in the fridge for about 2 weeks. Don't forget to save the whey; although it is no longer suitable for making cheese, there are still plenty of other uses for it.

Homemade Yogurt

To make your own yogurt, you will need a starter culture. If a friend has a particular variety you like, ask them for a few tablespoons of plain yogurt. Otherwise, you will have to get your culture from store-bought yogurt the first time. Different brands and varieties have their own tastes and textures, so pick the one you like best, whether that is a plain container of commercial yogurt or a more exotic Greek import. After your first batch, you can use a few tablespoons of your own yogurt to keep the process going.

Probably the most important and difficult part of making your own yogurt is the culturing time. During this time, you must keep your yogurt at a temperature between 90 and 110 degrees. Too hot or too cold and the cultures die. There are several ways you can maintain the temperature, including special yogurt-making machines. There are also plenty of options for making yogurt without special equipment, so choose whichever one is most convenient for you.

The first method is to use a crockpot. While you are heating the milk, plug your crockpot in, put the lid on it, and turn it on low to allow it to preheat. When the milk is ready to culture, you should unplug the crockpot, pour the hot milk into the crock, then replace the lid and wrap the whole thing in a towel. Notice that the crockpot is not actually plugged in while the milk

Fermentation Factor

is culturing. The preheating is enough to keep it warm for the necessary amount of time.

A second option is to use a heating pad and an appropriately sized bowl with a lid. Pour the milk into the bowl and set the heating pad on medium. After covering the bowl with a towel, you can either place the heating pad on a cutting board and then set the bowl on top of it, or you can wrap the heating pad around the bowl.

Your third option is to use a small insulated or styrofoam cooler. Pour the hot milk into one or several mason jars (small enough to fit inside the cooler), then fill the cooler with 110 degree water, close, and wrap with a towel. You will want to check the water temperature every couple of hours—if the water temperature begins to approach 90 degrees, remove about half the water and replace with more 110-degree water. Try to disturb your yogurt as little as possible while doing this.

Any of these methods will work, so choose the one that suits you best. All the rest of the steps are the same.

Equipment:
- Metal pot
- Candy thermometer
- Warming equipment of your choice
- Glass or plastic containers for storage

Ingredients:
- 1 quart milk
- 2 to 3 tablespoons plain yogurt
- Flavorings, if desired

1. Pour the milk into the metal pot and heat on the stove. Stir frequently until the milk reaches 185 degrees (when it begins to froth).

2. Remove the milk from the heat and allow it to cool to 110 degrees, stirring occasionally. Some people prefer to put the put in a sink full of cold water to speed up the process. This is completely fine.

3. Mix in the yogurt starter and transfer the milk mixture to whatever warming container you will be using.

Fermentation Factor

4. Set up whichever warming method you have chosen and allow the yogurt to culture for eight hours. Do not disturb the yogurt, as it is very temperamental during this stage.

5. After 8 hours of culturing, carefully move the containers of yogurt to the fridge. Try not to jostle the containers, as this can prevent the yogurt from setting completely. Allow to cool at least overnight (24 hours is preferable).

6. You can now mix in any flavorings you would like and move the yogurt to the storage container of your choice. Enjoy your homemade goodness within 2 weeks. Don't forget to set aside a few tablespoons of the fresh plain yogurt so you can make your next batch.

Making Whey

Plain yogurt is a great source of whey and probably the most accessible for the average person. If you don't have any of your own homemade yogurt and are in need of whey for a recipe, store-bought yogurt will still get you what you need.

Equipment:
- Colander or sieve
- Bowl that will hold your colander
- Cheesecloth

Ingredients:
- ½ cup gelatin-free plain yogurt

1. Line the colander or sieve with cheesecloth and place the colander into the bowl.

2. Pour the yogurt into the cheesecloth-lined colander. Gather up the edges and squeeze gently. About 4 tablespoons of whey should drip into the bowl.

CHAPTER 7

THIS AND THAT
Fermented or Not?

There are many foods that you've probably eaten that you never even realized were fermented. By contrast, there are also probably many things that you assumed were fermented when they actually weren't. This chapter will give you a few last true ferments, as well as address some of those deceptive pickled imposters.

This: True Fermented Foods

Did you know in addition to fruits, vegetables, grains, drinks, and dairy, you can also ferment meat? You might have even had some, if you've had a piece of genuine Italian salami or pepperoni. I'm not talking about the stuff that's packed with chemicals and irradiated (yes, irradiated) before it's delivered to your table. I'm talking about true fermented meats. There are very few places in the U.S. that still make these masterpieces, as the U.S.D.A. has made quick work of places that don't cook food the way they see fit (never mind that people have made these products for hundreds of years).

While homemade salami or pepperoni is a bit out of our scope without training from a master, there is another fermented meat you've probably had before and can make at home: corned beef.

Corned Beef

Adapted from a recipe found at the Nourished Kitchen, this corned beef is a great way to add the benefits of fermenting to your table. Enjoy it with a side of fresh sauerkraut and sourdough bread for a delicious and healthy meal.

Fermentation Factor

Equipment:
- Small mixing bowl
- Cheesecloth
- Kitchen twine
- Large glass or ceramic bowl with lid
- Heavy plate (able to fit inside the bowl)

Ingredients:
- 2 to 3 pound cut of beef
- ½ cup kosher salt
- 2 tablespoons black peppercorns
- 2 tablespoons mustard seeds
- 2 tablespoons coriander seeds
- 2 tablespoons hot red pepper flakes
- 2 tablespoons allspice berries
- 1 tablespoon ground mace
- 2 small cinnamon sticks, broken into pieces
- 2 to 4 bay leaves, crumbled
- 2 tablespoons whole cloves
- 1 tablespoon ground ginger
- 3 cloves garlic, chopped
- 2 cups fresh whey
- 2 cups celery juice
- Water, if necessary

1. Rinse the piece of beef, pat dry, and set aside.

2. Mix together all the seasonings except the whey and celery juice. Rub the spice mix into the meat, then roll the meat up in the cheesecloth and secure with the kitchen twine.

3. Place the meat in the large bowl and cover with whey and celery juice. If the liquid does not completely cover the meat, add a little water.

4. Use the plate to weigh the meat down and keep it below the brine, then cover and place in the fridge. Allow to ferment in the fridge for about a week, rotating the meat each day so that it cures evenly.

5. After the meat is finished fermenting, you can drain the brine and cook the corned beef in your favorite recipe. Enjoy!

Fermented Mayonnaise

What's a corned beef sandwich without a little mayo and mustard? You can also ferment your own condiments. Just make sure you follow the instructions carefully to ensure you get mayonnaise that sets properly.

Equipment:
- Mixing bowl
- Whisk or hand mixer (a blender also works well)
- Mason jar with lid

Ingredients:
- 3 egg yolks (room temperature)
- 1½ to 2 cups olive oil
- 3 to 5 tablespoons lemon juice or vinegar
- ½ teaspoon kosher salt
- ¼ teaspoon dry mustard
- 2 to 3 tablespoons whey

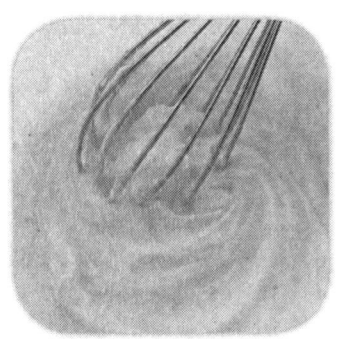

Photo courtesy of Wikimedia Commons by FotoosVanRobin from Netherlands

1. Make sure the egg yolks are at room temperature, then mix for at least 2 minutes.

2. Add the lemon juice or vinegar, salt, and mustard and mix for another minute.

3. While you are still whisking or blending, add the olive oil very slowly—1 or 2 drops at a time. If you go too fast, you'll end up with a runny mess. If the mayo becomes too thick, add a few more drops of lemon juice or vinegar.

4. Mix in the whey and transfer the mixture to the mason jar. Cover and allow to ferment at room temperature for about 6 to 8 hours, then move to the fridge. Enjoy!

Fermented Mustard

Enjoy experimenting with different types of mustard seeds until you find a flavor that suits you. Brown mustard seeds are hotter than yellow mustard. Also, if you like honey mustard, add in a tablespoon or two of honey to taste.

Equipment:
- Small mason jar with lid

Ingredients:
- ½ cup mustard powder or crushed mustard seeds
- 3 tablespoons whey
- 2 tablespoons water (if you want mustard that's a little tangier, try using lemon juice or vinegar in the place of the water)
- 1 teaspoon salt
- 1 tablespoon honey, if desired
- Additional spices (pepper, dill, etc.), if desired

Mix all ingredients together in the mason jar. Cover and allow to ferment for 3 days before moving to the fridge.

That: Pickled Doesn't Always Mean Fermented

As mentioned at the beginning of this book, just because something is pickled doesn't mean it is fermented. Most of the pickled products you buy at the grocery store are most certainly not fermented; the companies that produce them use vinegar and other acids to preserve the food. When you eat these sterile foods, you are missing out on the beneficial microorganisms present in the living food that is fermentation.

However, there is one issue that can come up with fermented foods. If you've been reading all the recipes, including the ingredients lists, you'll notice that many of these recipes include considerable quantities of both salt and sugar. If you have high blood pressure, heart disease, or diabetes, then these ingredients may be a concern to you. Because the levels of salt and sugar in the brines can be critical to whether your food become healthy ferments or rotten messes, it's not recommended that you tinker with those recipes. However, included here are a few refrigerator pickles that you can adapt to your health needs. If salt is an issue, simply reduce the amount or use a salt substitute. Sugar can be replaced with a commercial sugar substitute. As you give these a try though, keep in mind that these are not fermented foods but refrigerator pickles—they are meant to be stored in the cold and eaten within a week or two.

A note: as in the fermentation process, the acid in the vinegar used to make refrigerator pickles can become a bit temperamental when exposed to metal, possibly causing your brine to become cloudy. Try to use non-metal mixing bowls and containers whenever possible.

Refrigerator Dill Pickles

Equipment:
- Large mixing bowl
- Several mason jars with lids (size is up to you)
- Ladle

Ingredients:
- 12 pickling cucumbers (about 3 or 4 inches long)
- 2 cups water
- 1¾ cups white vinegar
- ½ cup sugar
- 1½ tablespoons salt
- 6 cloves garlic, chopped
- 1½ cups chopped fresh dill weed
- 1 tablespoon pickling spice (see Chapter 3)
- 1½ teaspoons dill seed
- ½ teaspoon red pepper flakes

1. If you'd like to have whole cucumber pickles, simply wash the produce before beginning. Otherwise, cut them into spears or slices.

2. Combine all the ingredients in the mixing bowl. Stir well and let stand at room temperature for 1 to 2 hours, or until the sugar and salt dissolve.

3. Remove the cucumbers from the liquid and pack into the mason jars. Ladle the brine over the pickles until they are covered and screw on lids.

4. Allow the pickles to sit in the fridge for several days to absorb the flavors from the brine before enjoying.

Sweet Refrigerator Pickles

Equipment:
- Large bowl
- Saucepan
- Ladle
- Several mason jars with lids (size is up to you)

Ingredients:
- 7 cups sliced cucumbers
- 1 large onion, thinly sliced
- 2 tablespoons salt
- Enough ice water to cover the cucumbers and onions
- 1 cup vinegar
- 2 cups sugar
- 1 teaspoon celery seed
- Pinch of allspice
- Pinch of red pepper flakes

1. Place the cucumbers and onions in the bowl and sprinkle the salt over them. Cover with ice water and allow to soak for 2 hours.

2. Drain the pickles and pack them into the jars.

3. Mix together the remaining ingredients in the saucepan and bring it to a boil. Ladle over the pickles until they are completely covered, then screw on lids and move to cold storage.

4. Allow the pickles to sit in the fridge for several days to absorb the flavors from the brine before enjoying.

Watermelon Rind Pickles

This Southern favorite is a great way to keep even the watermelon rinds from going to waste.

Equipment:
- Large pot
- Sieve or colander
- Large bowl with lid
- Heavy plate that will fit inside the bowl
- Large saucepan
- Ladle
- Several mason jars with lids (size is up to you)

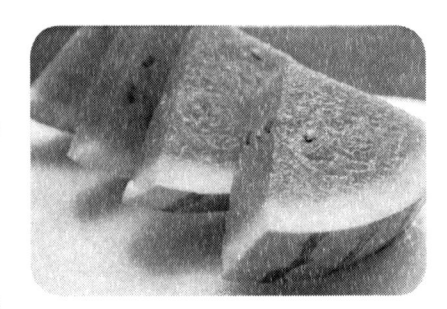

Ingredients:
- 4 pounds watermelon rind
- 8 cups water
- 8 teaspoons salt
- 2 cups sugar
- 1¼ cups apple cider vinegar
- ½ teaspoon pickling spice (see Chapter 3)
- ¼ teaspoon ground ginger
- 8 whole cloves
- 8 whole black peppercorns
- 2 cinnamon sticks, broken into small pieces
- ¼ teaspoon ground allspice

1. To prepare your watermelon rind, cut the watermelon pulp and the green skin away from the rind (leaving only the greenish-white center portion). Cut the remaining rind into pieces about ½ inch thick and 1 to 2 inches long.

Fermentation Factor

2. Combine the water and 2 tablespoons of the salt in the large pot and bring to boil. Add rind pieces and boil until tender (about 5 minutes). Strain and place in the large bowl.

3. Combine remaining ingredients in the saucepan. Bring to boil, stirring frequently. Pour over the watermelon rinds in the bowl. Use the plate to keep the watermelon pickles below the brine. Cover and refrigerate overnight.

4. Strain the syrup from the rinds and heat it to boiling in the saucepan, then pour back over the rinds. Recover the bowl and return it to the fridge overnight again.

5. Repeat step 4 one more time, then transfer pickles to mason jars and screw on lids. Store in the fridge.

Squash Pickles

Equipment:

- Large mixing bowl
- Saucepan
- Sieve or colander
- Ladle
- Several mason jars with lids (size is up to you)

Ingredients:

- 1 pound zucchini and/or summer squash
- 1 small yellow onion, thinly sliced
- 2 green peppers, sliced
- 2 tablespoons salt
- Enough ice water to cover the vegetables
- 2 cups cider vinegar
- 1 cup sugar
- 1 teaspoon pickling spice (see Chapter 3)
- 1½ teaspoons crushed mustard seeds

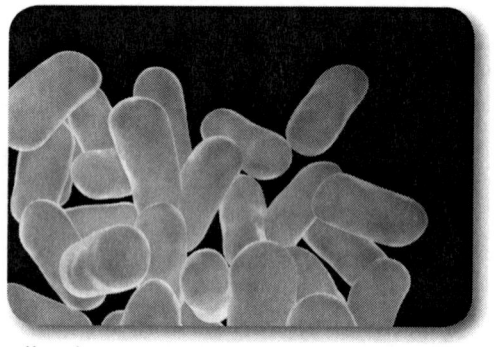

- 1 teaspoon ground turmeric

1. Wash and trim the squash, then slice, cube, or cut it into spears.

2. Combine the vegetables in the mixing bowl and sprinkle with the salt. Cover with the ice water and allow to soak for at least 1 hour.

3. While the vegetables are soaking, combine the remaining ingredients in the saucepan and simmer until the sugar is dissolved. Allow to cool to room temperature.

4. Drain the water off the squash and pack it into jars. Ladle the brine over the squash, making sure that the liquid completely covers all the vegetables. Screw on the lids and move to cold storage. Allow the pickles to soak for at least 1 or 2 days to absorb the flavors from the brine.

CHAPTER 8

How to Grow Your Own Food to Ferment

One important part of self-sufficient food production and storage is the ability to grow your own food. In this book, we've included the most commonly fermented foods and instructions about how to grow them in a home garden.

Cabbage

Companion Planting And Crop Rotation

Cabbage plants and other plants in the cabbage family (broccoli, kale, and Brussels sprouts) need to be rotated (in terms of growing location) to avoid soil fungal diseases. Never plant a cabbage plant or a plant in the cabbage family in the same place more than once every three to four years. That makes rotating this crop tricky! There are also early, mid-season and late varieties of cabbage. To enjoy cabbage for the longest period of time, plant some of each variety.

Cabbage grows well with artichokes, beets, bush beans, cucumbers, lettuce, peas, potatoes, and spinach. Keep cabbage away from basil, tomatoes, and strawberries.

Starting Seeds Indoors

Sow indoors ¼ inch deep in pots or flats about eight weeks before the last frost. Thin seedlings when they are 2" tall and transplant into individual pots.

Planting/Transplanting

Transplanting seedlings: Plant outside when a light frost is still possible. Plants can be transplanted 14-21 days before the average last frost or 90 days before the first frost in the fall.

Spacing: Plant outdoors 24 inches apart in rows 36 inches apart.

Soil type: The pH of soil in which cabbage grows is very important to avoiding diseases that strike plants in the cabbage family. Cabbage grows best with a soil pH of between 6.0-7.5. A pH of 7.2 or above helps keep clubroot from occurring. Adding lime raises the pH.

Sunlight: Cabbage needs full sun.

Seasonal Care

Watering: Water heavy early during the growing season and taper off to even moisture as the heads begin to form.

Feeding: Cabbage needs heavy nitrogen and potassium feeding.

Pollination: Because you eat the leaves of the cabbage plant, pollination is not important for formation of the food aspect of cabbage. For a cabbage to "bolt" and flower, you may need to cut an opening in the top of the cabbage head.

Special Requirements: The pH must be carefully monitored with cabbages. Cabbage plantings must be rotated to different areas of the garden, with cabbage plants and close relatives not occupying the same place more frequently than every three years.

Pests And Diseases Affecting Cabbage

Cabbage plants are affected by many pests and diseases.

Pests: Aphids, cabbage butterflies, cabbage loopers, cabbage maggots, cabbageworms, cutworms, diamondback moths, flea beetles, green worms, harlequin bugs, leaf miners, mites, moles, root flies, slugs, stink bugs, and weevils.

Diseases: Black leg, black rot, clubroot, damping off, fusarium wilt, leaf spot, pink rot, rhizoctonia, and yellows.

Harvesting For Food

Which part of the plant to harvest: The cabbage "head," which is actually tightly packed leaves.

When to harvest: Cabbage takes between 90-125 days to reach maturity, depending on the variety. You may harvest cabbage at any time during its growth by cutting off the main stalk at the ground level. Do not wait for a hard freeze to happen, though cabbage can be harvested through light freezes. Spring planted heads should be harvested before the outer leaves begin to turn yellow and wilt and the stem begins to elongate.

How to harvest: Cut the heads of any cabbage you wish to eat when the heads are full and firm. Keep the roots on the plant and wrap the cabbage in newspapers. Store in a cool, humid location.

How to prepare: Cabbage can be shredded and eaten raw, sautéed, boiled, or pickled and served.

Cabbage Storage

Storage conditions: Store cabbage at temperatures between 32-40 degrees Fahrenheit and at 80-90% humidity.

Harvesting Seed

Biennial. Cabbage will cross-pollinate with all other Brassica oleracea, (broccoli, kale, turnips) so isolate by one mile the during second year when going to seed or bag the tops of the plants to isolate. Do not harvest heads on plants you intend to save for seed. Carefully dig the plants and pot them in sand. Store plants between 32-40 degrees Fahrenheit. Plant back out in early spring and allow to bolt. You can cut an "X" in the top of the cabbage head to make it easier for the seed stalk to grow out of the head. Harvest seed pods when dry and clean by hand. The pods will ripen gradually from the bottom of the plant to the top of the stalk. The pods must come to maturity fully while still attached to the plant, or they will not germinate well the next year. Harvest each pod as it is ripe. If you want to save seeds from numerous types of Brassica, it is a good idea to alternate growing years for consumption and harvest.

Cucumbers

Companion Planting And Crop Rotation

Cucumbers are good garden bed mates with beans, cabbage, eggplant, kale, melon, peas, sunflowers, and tomatoes.

Starting Seeds Indoors

You can start cucumbers inside, but cucumbers don't like to have their roots disturbed. If you live in a cooler area, start cucumber seeds in peat pots so that they are not disturbed when you transplant them. Allow three weeks from planting seeds inside until moving outside.

Planting/Transplanting

Transplanting seedlings: If you start seeds inside, take care not to disturb the roots when you plant them outside. Plant outside when it is warm outside. Cucumbers like it hot!

Spacing: Place one plant every three feet. Cucumbers like to climb, so provide support on a trellis.

Soil type: These plants benefit from lots of compost and organic matter in the soil.

Sunlight: Cucumbers need full sun to produce fruit.

Seasonal Care

Watering: Cucumbers need lots of water. Keep the soil evenly moist.

Feeding: Feed cucumbers every three weeks with liquid fertilizer.

Pollination: In order to produce fruit, cucumbers have to be pollinated by insects. If insects are scarce, you can hand-pollinate by using a soft paintbrush to move pollen between flowers.

Pests And Diseases

Pests: Cucumbers are affected by aphids, cucumber beetles, cutworms, root-knot nematodes, and squash vine borers.

Fermentation Factor

Harvesting For Food

Which part of the plant to harvest: Harvest the fruits.

Cucumber Storage

Cucumbers should be sour pickled to keep for long periods outside of refrigeration.

Radishes

Companion Planting And Crop Rotation

Radishes are often planted as companion plants in the garden. They sprout and mature quickly, which means that they're out of the way by the time later crops need space. Many pests are deterred by the scent of radishes. Almost every plant benefits from radishes planted nearby.

Starting Seeds Indoors

Radishes are a root crop and should be planted directly into the garden, not transplanted.

Planting/Transplanting

Planting: Sow seeds directly into the garden when the temperature is at least 50 degrees or higher.

Spacing: Plant radish seeds three inches apart. Radishes are good germinators, so you don't have to sow more than you want to harvest. (Thinning radishes is annoying!)

Soil type: Radishes are widely adaptable to all soil types.

Sunlight: Radishes grow well in full sun, but can tolerate some morning shade.

Seasonal Care

Watering: Water regularly, keeping the soil evenly moist.

Feeding: Radishes benefit from one feeding during their short growing season.

Pollination: Radishes are pollinated by insects. If you want radishes to go to seed, leave a few in the ground to bloom. This is a good way to attract pollinating insects to your garden.

Special Requirements: Radishes are one of the easiest plants to grow.

Pests And Diseases

Radishes are affected by few pests and diseases.

Harvesting For Food

Which part of the plant to harvest: Pull up the entire plant and eat the root.

When to harvest: Radishes are ready to harvest 45-60 days after seeding.

Radish Storage

Store radishes in a cool, moist place.

Harvesting Seed

Radishes have to be allowed to "bolt" and flower in order to go to seed. Let the seed heads form on the plant. Just before they have started to dry and pop open, pull up the entire plant and hang it upside-down over newspaper. The seeds will dry and fall out onto the newspaper.

Beets

Companion Planting And Crop Rotation

Beets are cool season crops that can be sown early in the spring and then again in the early fall. Plant rows of beets every two weeks to have fresh beet root for a several-month period. In-between, the area where beets are planted can be planted with squash, cabbage-family plants, garlic, onions, and head lettuce. Keep beets away from field mustard and pole beans.

Starting Seeds Indoors

Beets grow best when started directly outside in the garden.

Planting/Transplanting

Starting seeds outdoors: Beets germinate best in temperatures of 50-85 degrees Fahrenheit. They should be sown in the spring 2-4 weeks before the average last frost date.

Spacing: Beets can have spotty germination rates, so you will want to sow them thickly and thin them once they have begun setting their first leaves. Beet greens are great additions to salads, so the thinned beets won't go to waste.

Soil type: Beets need soils high in organic matter, and a low nitrogen to phosphorous ratio. Excessive nitrogen causes over-production of leaves, and small roots.

Sunlight: Beets need full sun.

Seasonal Care

Watering: Maintain an even moisture level in the area where you grow beets.

Feeding: Every two weeks, feed with a balanced organic fertilizer. Once the beet leaves are 4-6 inches long, feed with a fertilizer lower in nitrogen.

Special Requirements: Watch the nitrogen amount in the soil.

Companion Plants: Beets grow well with plants in the cabbage family, beans, and head lettuce. Beets do not grow well with mustard plants.

Pests And Diseases Affecting Beets

Beets are fairly pest and disease free. They are sometimes affected by beet leafhoppers, carrot weevils, earwigs, garden webworms, leaf miners, mites, whiteflies and wireworms.

Downy mildew, leaf spot, rust and scab are diseases that sometimes affect beets.

Harvesting For Food

Which part of the plant to harvest: The entire beet plant is edible. The leaves can be eaten raw in salads when young, or sautéed at any size. Beet roots can be eaten raw, roasted, or pickled.

When to harvest: Harvest beets when young for salads. Roots can be harvested after 50-70 days, depending upon the variety.

How to harvest: Pull beets in the morning, before the leaves have lost water. Do not cut the tops off the beets until you are ready to cook them, as they can lose some of their nutrients by "bleeding" when they are cut up or the leaves are cut off.

How to prepare: Scrub the beet roots thoroughly but do not peel them unless absolutely necessary, in order to keep as much of the "juice" inside the beets. Beets can be grated raw into salads, roasted in the oven, and boiled and pureed for soup.

Beet Storage

Storage conditions: Beets will keep for two weeks when bunched together and kept in the refrigerator.

Corn

Companion Planting And Crop Rotation

Sweet corn tastes best when picked and immediately eaten. For this reason, it is a good idea to plant varieties that mature at different times, or sow successive plantings of the same variety 2-3 weeks apart. Plant corn for staggered harvest in four-foot square blocks. Corn is a heavy feeder, so alternate corn patches with beans every other growing season.

Corn grows well with beans, clover, peanuts, sweet potatoes, cucumbers, melons, pumpkins, squash and soybeans. Keep it away from tomatoes, as certain insects enjoy feeding on tomatoes and corn.

Starting Seeds Indoors

Unless you are growing corn in an area with cool, late springs, it is much better to start corn outside than inside. However, if you do live in an area with cool springs, you can sow corn in plug flats-- one kernel per plug, about 1 inch deep. Transplant when the corn has its first two true leaves.

Planting/Transplanting

Starting seeds outdoors: Sow seeds outdoors only after the danger of frost has passed. Corn will not germinate properly when the soil is still cold in the spring. Sow seeds 1 inch deep every 3-4 inches in rows 3-4 feet apart. Thin the seedlings to 8 inches apart after the plants come up. Corn should be planted in a 3-4 row block to ensure well filled-out ears.

Soil type: Corn will grow well in a wide range of soil pH. It does need high levels of nitrogen.

Sunlight: Corn needs full sun to grow and fruit.

Seasonal Care

Watering: Corn is fairly drought tolerant and will likely not need supplemental water unless there is severe drought.

Feeding: Corn is a heavy feeder. Make sure that soil has plenty of organic matter and compost added prior to planting. Side-dress with compost once a month during the growing season.

Pollination: Corn is wind pollinated and needs to be planted close together to pollinate well and set good, filled-out ears.

Special Requirements: If you want to save corn seeds, you will need to isolate varieties by at least 1,000 feet.

Pests And Diseases Affecting Corn

Pests: Aphids, birds, corn borers, corn earworms, corn maggots, corn rootworms, cucumber beetles, cutworms, earwigs, flea beetles, garden webworms, Japanese beetles, June beetles, leafhoppers, sap beetles, seedcorn maggots, thrips, webworms, white grubs, and wireworms.

Diseases: Bacterial wilt, mosaic viruses, rust, smut, southern corn leaf blight, and Stewart's wilt.

Harvesting For Food

What part of the plant to harvest: Corn "ears" are the fruit of the corn plant.

When to harvest: Harvest sweet corn when the silks have just barely started to turn brown. If you pierce a kernel with your thumbnail and "milk" shoots out, the corn is ready to eat. Pull the ear off the stalk and leave the stalk with an ear to ripen for seed.

How to prepare: Corn can be baked, grilled, microwaved, or boiled. It can be cooked on and off the cob.

Corn Storage

Storage conditions: Fresh corn should be eaten within three to four days of harvest. (It tastes best right off the stalk but will keep in the refrigerator for a few days.) You can also cut corn off the cob and freeze it.

Harvesting Seed

All corn varieties are wind-pollinated and will cross-pollinate with each other. Varieties should be hand-pollinated or isolated by 1 mile to ensure purity. Allow ears to dry on the plants, then harvest and shell. To preserve genetic diversity of seed corn, take ears from at least twenty five different plants to harvest.

Tomatoes

Companion Planting And Crop Rotation

Tomatoes like it hot. They are perfect to plant in areas where spring peas have been growing, along radishes, carrots, cabbage, kale, or scallions. Keep tomatoes away from corn, dill, fennel, and potatoes. The number of tomatoes you plant depends entirely upon how much you love tomatoes and how many ways you use them to cook. There are hundreds of varieties of tomatoes, each maturing at a different time and with a different size and color. For tomatoes all season long, plant varieties that mature at different times. Also consider two tomato plantings, one about four to six weeks after the first (if you have a long growing season). That will ensure that if some plants become "exhausted" or succumb to disease, you have younger plants to pick up the slack.

Starting Seeds Indoors

Temperature: Tomatoes are warm-weather plants. Using a heating mat under the seedling flat will keep plants compact, rather than leggy.

Time before last frost: Sow indoors ¼ inch deep in pots or flats 6 weeks before the last frost.

Care: Thin seedlings when 2 inches tall and transplant into individual pots. Remove all but the top four leaves and bury the rest of the plant in the soil.

Planting/Transplanting

Transplanting seedlings: Transplant outside when all danger of frost is past. Remove all but the top few leaves from the plant and bury the rest of the stem in the soil. This encourages deeper rooting and better growth.

Spacing: Plant outdoors 2 feet apart in rows 3 feet apart. Indeterminate vines will require support. (These are non-bush forming tomato plants.) Never underestimate the power of a tomato plant to overcome and destroy any sort of support system.

Soil type: Tomatoes do best in slightly acidic soil, with a pH of 5.8-7.0. They are heavy feeders and do well when compost or fertilizer is worked into the soil at least one week before planting.

Sunlight: Tomatoes need full sun to produce heavily. If planting tomatoes in a smaller garden, always plant them on the north side (northern hemisphere) or south side (southern hemisphere) so that they do not block the sunlight from other vegetables.

Seasonal Care

Watering: Tomatoes need even moisture throughout the growing season. A wet/dry/wet/dry watering situation will cause blossom end rot of the tomato fruits.

Feeding: Tomatoes are heavy feeders and do well in rich, organic soil. Keep high-nitrogen fertilizers away from the plant once it starts flowering, or the plant will produce lots of healthy green leaves but few fruits. During fruiting, a fertilizer that is higher in phosphorous will help with production of larger harvests.

Pollination: Tomatoes must be insect pollinated (or hand-pollinated, in the case of greenhouse-grown crops) in order to produce fruits.

Companion Plants: Cabbage, carrots, cucumbers, chives, marigolds, peas, and nasturtiums.

Special Requirements: Tomatoes are susceptible to a huge variety of pest and disease problems. One way to cut down on soil-borne diseases is to solarize the soil. Tomatoes are one of the few crops that will benefit from a clear plastic sheet mulch around them. They can tolerate the heat, and the sunlight will help sterilize the soil. Just make sure that the plants stay well watered.

Pests And Diseases Affecting Tomatoes

Tomatoes are plagued by an almost endless variety of pests and diseases.

Pest problems: Aphids, beet leafhoppers, cabbage loopers, Colorado potato beetles, corn borers, corn earworms, cucumber beetles, cutworms, flea beetles, gophers, Japanese beetles, lace bugs, leaf-footed bugs, mites, nematodes, slugs, snails, stinkbugs, tomato hornworms, and whiteflies.

Disease problems: Alternaria, anthracnose, bacterial canker, bacterial spot, bacterial wilt, botrytis fruit rot, curly top, damping off, fusarium wilt, nematode, soft rot, spotted wilt, sunscald, tobacco mosaic, and verticillium wilt.

Plant care disorders: Blossom end rot and sunscald.

Harvesting For Food

Which part of the plant to harvest: Fruits.

When to harvest: Harvest tomatoes for eating when they have reached their final color but the skin is still taut.

How to harvest: For tomatoes that will sit for a few days before consumption, it is better to harvest with a bit of the stem left on. The little green leaves round the top of the tomato are called the calyx, and it is perfectly fine to leave that on the fruit.

How to prepare: Tomatoes are healthy and great-tasting when eaten fresh, chopped in salads, added to pasta, used on sandwiches, layered between fresh mozzarella and basil leaves, or cooked in soups, stews and sauces.

Tomato Storage

Storage conditions: Fresh, ripe tomatoes keep for two or three days. Green tomatoes can keep for up to two weeks and will slowly turn a bright orange. (They just won't get any sweeter once they are picked.) Canning, drying, and freezing tomatoes also works well. Dried tomatoes need to be kept in a place with low humidity, or they will get moldy fast. Frozen tomatoes should be used for cooked dishes, as the freezing destroys the cell walls and makes them mushy. Tomatoes are one of the few things that can be canned without a pressure canner, as long as there is little more than tomatoes and lemon juice in the jar. Tomatoes stay acidic enough to keep botulism out of the canned goods.

Harvesting Seed

Tomatoes, for the most part, do not cross-pollinate with each other, so you do not need to worry about isolating tomato varieties in order to save seed. To harvest the seed, simply save several tomatoes (label them by variety), and scoop the seeds out of the fruit. In order to remove the gelatinous coating from around the seed you will have to "ferment the seed." In order to keep your tomato varieties separate, you will need to ferment them in separate dishes. Here's how you ferment the seed:

Squeeze the seeds and gel coverings into a plastic container.

Set them in a garage or somewhere away from the kitchen for about three days.

Once a layer of mold has completely covered the seed mixture, bring the container back inside and pour out into a strainer.

Wash the seeds by running water through the strainer and rubbing the seeds gently against the strainer.

Drying seeds: Use direct sunlight. Do not dry on paper products because they will stick to the seeds.

Storing seeds: Seeds needs to be completely dry and stored in a cool, dry, dark place. Do not dry the seeds in an oven.

Fermentation Factor

Beans

Companion Planting And Crop Rotation

Snap beans are warm-season crops. Prior to planting beans, you can grow carrots, radish, peas, and strawberries. Keep basil, garlic, fennel, onions, beets, and cabbage away from snap beans. The number of plants depends upon whether you plan to freeze or can any of your harvest. Snap bean plants need to be harvested almost every day when they are producing heavily. Plant at least 10 plants per person for in-season eating, and 20 per person if you plan to preserve.

Starting Seeds Indoors

Beans grow best when started directly from seed outside. They germinate quickly when the soil temperature is warm.

Planting/Transplanting

Starting seeds outdoors: Sow seeds directly outside when the danger of last frost has passed.

Spacing: Plant seeds 1 inch deep and 2 inches apart in rows. Pole beans or climbing beans will need support.

Soil type: Beans can grow in almost any type of soil because they grow symbiotically with nitrogen fixing organisms.

Sunlight: Beans need full sun to grow and produce well.

Seasonal Care

Watering: Beans are fairly drought tolerant but will produce juicier beans for eating fresh when they are given ample water.

Feeding: Beans can fix their own nitrogen but need to have enough potassium in the soil to grow well.

Pollination: If you are growing two different types of beans for drying, plant them on opposite sides of the garden, as they will cross-pollinate with each other.

Special Requirements: Beans are one of the easier vegetables to grow. They do need to be harvested every day during the height of their production season.

Harvesting For Food

Which part of the plant to harvest: The fruit (the bean pod is the fruit of the bean).

When to harvest: Many beans can be eaten young and green, if necessary. The main purpose of growing beans is to preserve the beans through drying to rehydrate later in the winter or save for another time. Wait until the beans are dry on their stalks to harvest. For fresh beans, harvest when they are still young and tender.

How to harvest: Break the young beans off the plant with the end of their stem still attached. Harvest dry beans when the entire pod is dry and about to shatter.

How to prepare: Fresh beans can be grilled, "broken" (both ends broken off) and boiled, or steamed. Dry beans can be saved and rehydrated for soups, stews, and any number of other dishes.

Bean Storage

Fresh beans keep for a few days when refrigerated. Dry beans must be kept very dry in order not to sprout or mold.

Harvesting Seed

Bean seed pods that are left on the vine to dry can shatter upon touch. Carefully close your hand around the pod, pull the entire pod off the plant, and place it in your harvest container.

Preserving Seed

Cleaning Seeds: Winnow the chaff away from the beans. Sift the small, dry plant parts away from the dry beans.

Drying Seeds: To kill bean weevils, freeze the seeds for at least three days before saving them in an airtight container. You will know that seeds are dry enough to freeze if they shatter when hit with a hammer.

Storing Seed: It is always best to save seed from plants that ripen first and are free from disease. Harvest seed pods when completely dry, crush in a cloth or burlap sack, and winnow the seeds from the chaff. If you have to pull your crop before it has completely dried, hang the plants upside down in a burlap bag and let them dry. As soon as they are dry, store them in a cool, dry, dark location.

Quinoa

Companion Planting And Crop Rotation

Quinoa is a cool-season plant. It can be planted with all cool-season vegetable crops. Plant ten plants per person for preserving.

Starting Seeds Indoors

Quinoa is a cool-weather crop that does best when directly seeded in the garden.

Planting/Transplanting

Starting seeds outdoors: Sow seeds directly into the garden when soil temperatures are about 60 degrees Fahrenheit.

Spacing: Quinoa can be thinned and the young leaves eaten steamed or in salads, like spinach, so spacing is not a concern with these plants.

Soil type: Does well in fertile, well-drained soils and is extremely drought tolerant.

Sunlight: Needs full sun.

Seasonal Care

Watering: Needs very little supplemental water and is a drought-tolerant plant.

Feeding: These plants require little supplemental food.

Pollination: These plants will out-cross with other closely related plants (such as lamb's quarters). If you want to save seed, you need to do a thorough check of your surrounding area to remove plants that may cross with the quinoa.

Special Requirements: This plant grows well in areas with warm days and cool nights and is well-adapted to high-altitude growing conditions.

Pests And Diseases Affecting Quinoa

Occasionally subject to insect damage but relatively pest-free.

Harvesting for Food

What part of the plant to harvest: Young leaves can be picked and eaten like spinach or swiss chard. The "grain" aspect of quinoa are the seeds, which should be harvested from the plant when they are fully dried.

When to harvest: Harvest the leaves when they are young and relatively small. Harvest the seeds when they are drying on the plant.

How to harvest: Harvest by pulling the seeds off the plant by hand.

How to prepare: To prepare quinoa seeds for eating, you need to rinse them and taste them to make sure that they do not taste bitter. There are compounds covering quinoa seeds called "saponins" that need to be removed before eating. Boil one part seeds with two parts water for about 15-25 minutes, or until tender. Toss with fresh vegetables, nuts, fruit, spices, etc. This is a versatile food that can take the place of many grains and is tasty prepared sweet or savory.

Quinoa Storage

Storage conditions: The seeds must be dry before they are put into storage and should be dried at the temperature at which they will be stored. They should be stored in cool (not cold), dry locations.

Turnips

Companion Planting And Crop Rotation

Turnips grow best as other vegetables are withering. They can be planted in July or August for a winter crop. Plant turnips near onions and peas for the best results. Turnips can be grown for the roots, leaves, or both. If you plan to mainly make use of the leaves, you can sow them thickly and refrain from thinning them throughout the growing season.

Starting Seeds Indoors

Turnips don't transplant well from inside to the garden, so start the seeds outside. Turnip seedlings are too small and delicate to be handled very much.

Planting/Transplanting

Transplanting Seedlings: Start seeds outside about 70 days before the first frost.

Spacing: Sow the seeds one half inch deep and an inch apart. Rows should be one foot apart.

Soil type: Turnips prefer soil that is slightly acidic with a pH between 6.5 and 7.0. The soil should be loose and well-draining.

Sunlight: Turnips tolerate partial shade.

Seasonal Care

Watering: Turnips need at least an inch of water per week. The crucial period for watering is when the roots are just developing. Without adequate water, the roots can become tough and bitter.

Feeding: Fertilizer is not necessary for turnips. They grow quickly, but should be planted in soil with rich, organic matter.

Pollination: Turnips are cross-pollinators. The job is done by bees and other insects.

Special Requirements: Turnips are easy to grow and do not require significant special maintenance.

Pests and Diseases

Pests: Aphids and flea beetles will eat turnip greens. This can be avoided by covering the row. Because of the edible root, maggots and worms can be an issue. You can avoid them by using diatomaceous soil or by introducing nematodes or the rove beetle to the soil.

Diseases: Diseases such as white blister, mildew, and clubroot are possible, but unlikely. To avoid disease, you can rotate the location in which you plant the turnips year by year.

Harvesting For Food

Which part of the plant to harvest: Both the leaves and the root are edible.

When to harvest: The greens can be harvested after they are four inches tall. They will grow back. The roots are best at two to three inches. Turnips grow quickly, so the roots will be ready within four to five weeks. For turnips planted in the fall, the roots can be harvested at any time in the winter. They stop growing, but get sweeter the longer they stay in the ground.

How to harvest: Break off greens without damaging the root. Pull roots up when they are still small.

How to prepare: Fresh, small roots can be peeled and eaten raw. They can also be boiled or roasted. Young greens can be eaten raw, while older greens are best sautéed or steamed.

Turnip Storage

Greens should be eaten immediately. The roots can be stored for months in a cool, dark place. Remove the greens before storing roots.

Harvesting Seed

Seeds can be harvested from turnips that are not going to be trimmed and eaten. Wait for seed pods to dry on the plant.

Preserving Seed

Cleaning Seeds: Smash seed pods to remove seeds. Winnow the chaff.

Drying Seeds: Allow seeds to dry thoroughly before storing. They are dry when they break, rather than bend when struck.

Storing Seed: Seeds must be stored in a dry, cool place. Keep them in an airtight container like a glass jar.

Garlic

Companion Planting And Crop Rotation

This is a cool-season crop that you will plant in the fall. After harvesting garlic in the spring, you can plant beets, cabbage, celery, lettuce, and tomatoes. Keep beans and peas away from the garlic patch. Plant at least 20 starts per person.

Planting/Transplanting

Starting outdoors: It is most common for homeowners to grow garlic from garlic cloves. You can generally order them or pick them up at local garden centers. They grow best during cool weather and should be planted in the fall, with enough time to grow roots before the ground freezes.

Spacing: Plant bulbs two inches deep and about three or four inches apart.

Soil type: Garlic is a light feeder and does well when planted in rich, fertile soils.

Sunlight: Garlic needs full sun to develop well.

Seasonal Care

Watering: As a bulb, garlic does not need any supplementary water. Too much water will cause the bulb to rot.

Feeding: Doesn't need much fertilizer. Will grow well in compost-rich soils or when fed with an organic, seaweed-based fertilizer.

Special Requirements: To protect winter plantings from extreme temperatures, add four inches of mulch on top of the area where the garlic is planted.

Pests And Diseases Affecting Garlic

Garlic helps keep pests away from other plants and is not susceptible to many itself. Nematodes, botrytis and root rot are the only problems this plant must deal with. Controlling water helps to prevent problems with rot.

Harvesting for Food

Which part of the plant to harvest: The bulb is actually modified leaves.

When to harvest: Garlic can be harvested in the fall if planted in the spring, or in the spring if planted in the fall.

How to harvest: Allow the green tops to turn brown and fall over. Then dig (do not pull) up and allow to sit and dry on the top of the soil for a few days.

How to prepare: Garlic is good raw, roasted, or added to any cooked dish. It can be minced, crushed, and prepared with almost anything.

Garlic Storage

Storage conditions: Garlic must be stored in dry locations. You can braid the tops together to create a chain of garlic that will keep through the winter. Make sure to store without bruising to prevent rot.

HERBS

Coriander

Companion Planting And Crop Rotation

Coriander is a versatile and tasty herb. The plant itself is often called cilantro, while the seeds are referred to as coriander in cooking terms. It can be grown in a container or in the garden and thrives in sunny and cool weather. Coriander plants grow to be up to three feet in height and produce small, white flowers in the second year of growth. Coriander can be planted near tomatoes, potatoes, and eggplants to keep pests away. To help the coriander grow well, plant it near beans, peas, or anise.

Starting Seeds Indoors

Coriander seeds can be started indoors, but it is best to start outside. Starting them indoors can possibly lead to bolting.

Planting/Transplanting

Transplanting Seedlings: Sow the seeds outside just after the last frost.

Spacing: Place seeds in soil ¼ to ½ inch deep. Thin seedlings to keep them eight to ten inches apart, and keep rows at 12 to 15 inches apart.

Soil type: Soil should be acidic with a pH between 6.0 to 6.7. It should be mixed with good quality compost before starting seeds.

Sunlight: Coriander prefers full sun, but will tolerate a little bit of shade.

Seasonal Care

Watering: Keep coriander plants moist. You should never let them dry out entirely. Overhead watering can restrict the number of seeds produced, so water the plants at soil level.

Feeding: Using compost is adequate for coriander. You don't need to add fertilizer.

Pollination: Coriander self-pollinates. To allow it to do so, you need to allow the seed tops to grow.

Special Requirements: Coriander is easy to grow. It seeds very quickly, so if you want to keep using it throughout the growing season, you need to continually plant it.

Pests and Diseases

Pests: Coriander has no real pests. Most herbs repel insects and other critters.

Diseases: Diseases in coriander are also rare, although powdery mildew is possible.

Harvesting For Food

Which part of the plant to harvest: The leaves can be harvested to be eaten fresh. The seeds can be eaten as well.

When to harvest: The leaves of the coriander plant can be pinched off and used as needed once the plant is six inches tall. Leaves can be harvested throughout the growing season before the plant goes to seed. The seeds can be harvested about two to three weeks after the plant flowers. They should be light brown in color.

How to harvest: Break off leaves as needed. Remove from the tops of the stems to keep the plant from going to seed. For harvesting seeds, collect the entire flower and bundle them together. Put the bundle upside down in a paper bag and store in a dry place. The seeds will fall to the bottom of the bag when they are ready.

How to prepare: The fresh leaves can be eaten raw. The seeds can be ground into a powder and put into many dishes. They can also be roasted before grinding.

Coriander Storage

The leaves will lose flavor very quickly, so they should be used immediately. The seeds can be stored for several months in an airtight container in a cool, dry place.

Fermentation Factor

Harvesting Seed

Harvest seeds for future plantings in the same way in which you would harvest them for eating.

Preserving Seed

Cleaning Seeds: Simply let the seeds fall from the flowers as they dry in the paper bag.

Drying Seeds: The seeds will dry thoroughly in the paper bag.

Storing Seed: Store the seeds in an airtight container and keep them in a cool, dry place.

Dill

Companion Planting And Crop Rotation

Dill is an herb that makes a nice addition to many summer dishes. Although it can be eaten dried, the delicate flavor is best experienced fresh. Dill grows to a height of about three feet and produces small, yellow-green flowers, which attract beneficial insects to your garden. For that reason, it is beneficial to other plants to have dill nearby. Plants that should not be companions with dill include carrots and tomatoes.

Starting Seeds Indoors

Dill should be started from seed directly outdoors. Sow them just after the last frost.

Planting/Transplanting

Transplanting Seedlings: Sow the seeds outside just after the last frost.

Spacing: Space seeds outside to a depth of about ¼ inch. Sow in rows that are 18 to 24 inches apart. After about three or four weeks, thin the small plants so that they are 12 to 18 inches apart. Sow every few weeks for a continual supply of fresh dill.

Soil type: Dill grows best in rich, loose soil that is well-drained. The pH of the soil should be between 5.0 and 7.0.

Sunlight: Dill flourishes in full sunlight.

Seasonal Care

Watering: New dill plants need to be watered up to two times per week, enough to keep the soil moist. When the plants have grown to full height, they require very little watering because of the long tap root. Only water when conditions are very dry.

Feeding: Fertilizer is not necessary for growing dill as long as the soil is rich in organic matter.

Pollination: Dill, like other herbs self-pollinates. It also attracts pollinators like butterflies and bees. You can plant it near fruit trees and other cross-pollinators for this reason.

Special Requirements: Dill is fairly easy to grow, but you will need to stake it if your garden is in a windy spot. Its tall, delicate stalks can be broken otherwise. You can also consider growing a dwarf variety.

Pests and Diseases

Pests: Dill attracts several beneficial insects. The larvae of the black swallow butterfly eats dill. To keep them from destroying all of your dill plants, move any that show up to a single plant.

Diseases: There are few diseases that strike dill, but it may get powdery mildew.

Harvesting For Food

Which part of the plant to harvest: Dill leaves are eaten fresh. The seeds can be harvested for future plantings, or for use in pickling.

When to harvest: Dill leaves can be harvested approximately eight weeks after planting. Seeds are harvested from blooming dill plants, which occurs in late summer.

How to harvest: To harvest the leaves, pinch off close to the stem. Pick off the seed heads after they have turned a light brown color.

How to prepare: Leaves of the dill plant are best eaten fresh. They can also be dried or frozen and used later. The seeds need to be thoroughly dried and then can be used to add flavor to pickling vinegar.

Dill Storage

The leaves are best used immediately. They can also be dried and stored in an airtight container. They can also be frozen and stored in a plastic bag in the freezer. Dried seeds to be used for planting should be kept in an airtight container in a cool, dry place. If you plan to use them for pickling, store them in vinegar. They will become more flavorful the longer they stay in the vinegar.

Harvesting Seed

To harvest dill seeds, remove the seed heads from the plant and dry them in a paper bag.

Preserving Seed

Cleaning Seeds: Separate the seeds from the chaff by using a fan to blow off the light chaff.

Drying Seeds: The seeds will dry thoroughly in the paper bag.

Storing Seed: Store the seeds in an airtight container and keep them in a cool, dry place for reseeding. For pickling, store in vinegar.

Fennel

Companion Planting And Crop Rotation

Fennel is both an herb and a vegetable. You can eat the seeds, flowers, leaves, and the bulb or root. The plant has a delicate anise flavor and has long been used to aid digestion and to freshen breath. Fennel is a tall, feathery plant, similar in appearance to dill. It should not be grown near any other plants. It can cause many vegetables and herbs to bolt or to stop growing. Fennel can even cause other plants to die.

Starting Seeds Indoors

Fennel seeds are best started outside.

Planting/Transplanting

Transplanting Seedlings: Sow fennel seeds directly outside four to six weeks before the last frost.

Spacing: Sow the seeds 12 inches from each other and do a depth of ¼ inch. Rows need to be three feet apart.

Soil type: Soil for growing fennel needs to be well-drained and light.

Sunlight: Fennel needs full sunlight.

Seasonal Care

Watering: Water the soil lightly until the plant appears. Then, water once a week, as needed. It will not need much watering, as fennel has a long tap root.

Feeding: Enrich the soil every year with compost. Don't use fertilizer on fennel. A well-fed plant loses some of its flavor.

Pollination: To avoid cross-pollinating with other plants, bolting, and slow growth, grow fennel away from the rest of the garden. It is helped by being near dill, but this is not beneficial for the dill plants.

Special Requirements: Fennel can grow to three to four feet tall and needs to be staked. It also needs very well-drained soil.

Pests and Diseases

Pests: Pests that may attack fennel include carrot fly, slugs, caterpillars, cutworms, and carrot willow aphids.

Diseases: Fennel is susceptible to leaf blight and stem rot. Well-drained soil is a must to avoid rot.

Harvesting For Food

Which part of the plant to harvest: Every part of fennel is edible. You can harvest the leaves, bulb, flowers, and seeds.

When to harvest: The leaves can be picked as needed throughout the growing season. If you are planning to use the bulb, pull it up just before the plant flowers. If you are interested in harvesting the seeds, do so when they have turned brown. This will be in late August.

How to harvest: Pinch off leaves or flowers to use fresh as needed. To harvest the bulb, pull the entire plant from the ground. The leaves can be dried and stored. To harvest the seeds, cut off the stalk.

How to prepare: Eat leaves and flowers fresh. Leaves can also be dried, but are better when used immediately. The bulb can be eaten raw or cooked by roasting or sautéing. The seeds can be eaten once they are dried.

Fennel Storage

Use leaves immediately or dry and store in an airtight container. They can also be frozen. The bulb can be stored for a few days in the refrigerator. The seeds should be stored in an airtight container in a cool and dry location.

Harvesting Seed

To harvest fennel seeds, cut off the talk when the seeds are brown, but are not yet falling off. Hang the stalks to let the seeds dry completely.

Fermentation Factor

Preserving Seed

Cleaning Seeds: Remove the chaff by slapping the stalks with dried seeds against a hard surface.

Drying Seeds: Dry seeds by hanging the stalks in cheesecloth. They should dry within two weeks.

Storing Seed: Store the seeds in an airtight container and keep them in a cool, dry place for reseeding. For pickling, store in vinegar.

Sage

Companion Planting And Crop Rotation

Sage is a popular herb that can be used to flavor many different foods. It is a woody, evergreen, perennial plant, although it needs to be replaced every few years for best results. There are many different varieties of the sage plant including those that are best for eating and others that are great ornamental plants. In the garden, sage helps carrots by repelling carrot fly, but should not be planted near cucumbers. Near lettuce, sage can help deter slugs. Sage and rosemary are beneficial to each other and can be grown together.

Starting Seeds Indoors

Sage should be started from the seed outdoors rather than indoors for transplanting. Sow outside about six to ten weeks before the final frost.

Planting/Transplanting

Transplanting Seedlings: Sow seeds directly in the garden. Sage can also be propagated from cuttings.

Spacing: Place seeds in soil to a depth of one eighth of an inch. They can be scattered liberally, but when the seedlings appear, thin them so that they are two feet apart.

Soil type: Sage requires well-drained, rich soil.

Sunlight: Sage does best in full sun.

Seasonal Care

Watering: Keep the young sage plants well-watered until they have established themselves and have begun to grow well.

Feeding: Fertilizer is not necessary for sage, but the soil should be rich with organic compost.

Pollination: Sage, like most herbs attract pollinators to your garden. It will pollinate itself if you allow it to flower.

Special Requirements: As a woody, evergreen plant, sage needs to be pruned each year. Prune off the woodiest pieces in the spring as well as the flower stalks. You should stop harvesting sage in the early fall to allow the plant to harden for winter. After just a few years, the sage plants will become very woody and unproductive. They will need to be replaced.

Pests and Diseases

Pests: Spider mites, thrips, spittlebugs, whiteflies, and aphids can all present problems for sage.

Diseases: Sage may be susceptible to powdery mildew. Keep plants at an appropriate distance from each other to avoid this.

Harvesting For Food

Which part of the plant to harvest: The leaves of the sage plant are used to flavor foods and can be used fresh or dried.

When to harvest: Pick the leaves before the plant flowers. To get the seeds for further plantings, wait until after the plant has flowered. Do not over harvest the plant in its first year or it may not become well-established.

How to harvest: Pick leaves as needed throughout the growing season, before flowering. Harvest seeds by picking off the bell-shaped seed pods.

How to prepare: Sage leaves are excellent fresh and dried. They can be used in a variety of dishes. Do dry, hang the leaves in a shady, dry place. The leaves can also be frozen. Freeze in ice cube trays with water.

Sage Storage

Store frozen leaves in the freezer for several months. Dried leaves will keep in an airtight container, such as a jar for many months as well.

Harvesting Seed

To harvest sage seeds, pick the seed pods after the plant has flowered. The pods are shaped like bells and are ready when they are a dark brownish gray color.

Fermentation Factor

Preserving Seed

Cleaning Seeds: Simply let the seeds roll out of the mature seed pod.

Drying Seeds: After removing the seed pods from the sage plant, leave them to dry for a couple of days in a warm spot. There will only be a few seeds in each pod.

Storing Seed: Store sage seeds for future plantings in a closed container in a dry and cool location.

Thyme

Companion Planting And Crop Rotation

Thyme is a diverse and popular plant for gardening. There are over fifty varieties of thyme plants that can be used for culinary purposes or for decorative gardening. Thyme is easy to grow and does well indoors and in containers. Thyme's aroma helps to repel pests from the garden, especially white flies when planted near cabbage or broccoli. Keep thyme away from fast-growing herbs such as mint and tarragon. They can stunt the growth of a thyme plant.

Starting Seeds Indoors

Sage seeds can be started indoors or outdoors. If starting outdoors, sow seeds six to ten weeks prior to the final frost.

Planting/Transplanting

Transplanting Seedlings: If starting seeds inside, let them grow for eight to ten weeks. Transplant outdoors when the danger of frost is past. Thyme also does well when grown from cuttings.

Spacing: Place seeds in soil between six and twelve inches apart.

Soil type: Thyme grows best in soil that is light, well-drained, and warm. The pH of the soil can be anywhere between 6.0 and 8.0.

Sunlight: Thyme prefers to grow in full sunlight.

Seasonal Care

Watering: Young thyme plants should be watered well until they are established. After that, they prefer the soil to be fairly dry.

Feeding: You do not need to fertilize the soil for growing thyme.

Pollination: Thyme attracts pollinators like bees and butterflies. If allowed to flower, it will self-pollinate.

Special Requirements: Thyme is very easy to grow and requires minimal maintenance. You should, however, trim thyme plants a little bit during the spring and summer. This helps to keep it growing nicely and prevents too much woody growth.

Pests and Diseases

Pests: Not too many pests will bother thyme, although spider mites may be an issue.

Diseases: Root rot may occur if you over water your thyme plants. Keep the soil more dry than wet and be sure it is well-drained.

Harvesting For Food

Which part of the plant to harvest: The leaves of the thyme plant are used for culinary purposes. They can be used fresh, dried, or frozen.

When to harvest: Harvest leaves from your thyme plant throughout the growing season, with the final harvest in October. The best time to harvest the leaves for the purpose of drying is just before the plants flower.

How to harvest: You can pick individual leaves, but it is easier to pull of entire stems. The leaves are very small. When you have a stem, slide your fingers down it to release the leaves.

How to prepare: Use thyme leaves fresh in many dishes. To dry leaves, pick sprigs in the fall and hang them upside down in bunches. You can also freeze the leaves.

Thyme Storage

Frozen and dried leaves will last for many months if stored properly. Keep dried thyme in a n airtight container and store in a cool, dark, and dry place.

Harvesting Seed

Harvesting thyme seeds is not easy because they are so small. Thyme plants are best propagated by cuttings or root divisions. If you wish to collect the seeds, however, wait for the flowers to develop and then shake the seeds into a bag.

Fermentation Factor

Preserving Seed

Cleaning Seeds: Winnow the seeds from the chaff.

Drying Seeds: Allow the thyme seeds to dry thoroughly before storing them. They need to be dried in a warm, dry location.

Storing Seed: Store the completely dried thyme seeds in an airtight container. A glass jar with a lid works well. The seeds will keep in a cool, dry location for up to two years.

APPENDIX
Additional Resources

The sky (okay, the Internet) is quite literally the limit when it comes to resources for fermenting your own foods. There are dozens of recipes that didn't make the cut for this book, along with suggestions for modifying recipes to meet your individual tastes. Here are a few books and websites that I found to be the most helpful.

Wild Fermentation

Mentioned previously in this book, Sandor Ellix Katz's Wild Fermentation from Chelsea Green Publishing Company is a great book if you are interested in more fermentation recipes. In addition, he maintains a website (www.wildfermentation.com) that include more resources and forums to discuss your questions and ideas with other fermenters out there.

Nourished Kitchen

This great site has wonderful resources, including recipes, message boards, and cooking classes. It is easy to navigate and simple to understand. Jenny's advice is well worth considering. Check it out at www.nourishedkitchen.com.

Pickl-It

The Pickl-It is a unique device designed to combine all the tasks of fermenting into one easy-to-use contraption. If you are gadget nut, you may want to look into their offerings, but even if you aren't, they include lots of great recipes that you can try, even without a Pickl-It. Their online store is located at www.pickl-it.com.

Ball Blue Book

The "bible" of canning for decades, no kitchen is complete without this reference. Included in it are additional recipes and ideas for non-fermented pickles, along with guidelines for canning your pickles (whether they are fermented or not). Check out the website at www.freshpreserving.com.

Artisan Bread in Five Minutes a Day

This is quite possibly the best cookbook ever written. Although the recipes are written with store-bought yeasts in mind, a little creativity and practice could easily adapt them for use with your own home-dried yeast or liquid starter. The ease of the methods in this book will make you wonder why anyone would ever by store-bought bread. Learn more at www.artisanbreadinfive.com.